handbook of public communication of science and technology

Communicating science and technology has become a priority of many research and policy institutions, a concern of many other private and public bodies, and an established subject of training and education. In the past few decades the field has developed and expanded significantly, not only in terms of professional practice, but also in terms of research and reflection. As well as introducing the main issues, arenas and professional perspectives involved, this unique volume couples an overview of the problems faced by practitioners with a thorough review of relevant literature and research.

The handbook format makes this a student-friendly resource, but its breadth of scope and expert contributors mean it is also ideal for practitioners and professionals working in the field. Combining the contributions of different disciplines and the perspectives of different geographical and cultural contexts, this original text provides an interdisciplinary as well as a global approach to public communication of science and technology. Contributors include mass communication scholars, sociologists, discourse analysts, public relations practitioners, science journalists, and others. It is a valuable resource for students, practitioners and professionals in the fields of media and journalism, sociology, history of science, and science and technology.

Massimiano Bucchi is Professor of Sociology of Scienlice at the University of Trento, Italy. He is author of several books including *Science and the Media* (Routledge, 1998) and *Science in Society* (Routledge, 2004), and is co-editor of *Journalism, Science and Society Relations* (with M. Bauer, Routledge, 2007).

Brian Trench is Senior Lecturer and former Head of the School of Communications, Dublin City University, Ireland. He researches models of science communication, public representations of science and emerging technologies in the knowledge society, and social uses of the internet.

Massimiano Bucchi and Brian Trench are members of the scientific committee of the international Public Communication of Science and Technology (PCST) network.

handbook of public communication of science and technology

edited by
massimiano bucchi
brian trench

LONDON AND NEW YORK

First published 2008
by Routledge
2 Park Square, Milton Park, Abingdon, Oxon OX14 4RN

Simultaneously published in the USA and Canada
by Routledge
270 Madison Avenue, New York, NY 10016

Transferred to Digital Printing 2008

Routledge is an imprint of the Taylor & Francis Group, an informa business

Typeset in Bembo and Helvetica by
Taylor & Francis Books
Printed and bound in Great Britain by
TJI Digital, Padstow, Cornwall

British Library Cataloguing in Publication Data
A catalogue record for this book is available from the British Library

Library of Congress Cataloging in Publication Data
Handbook of public communication of science and technology / edited by
Massimiano Bucchi and Brian Trench.
p. cm.
Includes bibliographical references.
(electronic) 1. Communication in science–Handbooks, manuals, etc. 2.
Technical writing–Handbooks, manuals, etc. I. Bucchi, Massimiano, 1970–
II. Trench, Brian.
Q223.H344 2008
501′.4–dc22 2007042978

ISBN 978-0-415-38617-3 (hbk)
ISBN 978-0-203-92824-0 (ebk)

Contents

Illustrations

Figure

Tables

Boxes

Contributors

Martin W. Bauer is Reader in Social Psychology and Methodology at the London School of Economics, UK. He directs the programme 'Social and Public Communication', and researches science communication and the public understanding of science in comparative perspective, and the role of resistance in socio-technical processes. Books include *Genomics and Society* (edited with G. Gaskell, Earthscan, 2006), *Journalism, Science and Society* (edited with M. Bucchi, Routledge, 2007) and *Atoms, Computers and Genes – Public Resistance and Socio-Technical Responses* (Routledge, 2008).

Rick E. Borchelt is Lecturer in Science Policy and Politics at The Johns Hopkins University in Baltimore, Maryland, USA. He has served as Director of Communications for the US Department of Energy's Office of Science, where he implemented a strategic communications plan for public outreach on the Department's science portfolio. He has practised science and technology public affairs at the University of Maryland, the National Academy of Sciences (and its sister organisations, the Institute of Medicine and National Academy of Engineering), as press secretary to the House of Representatives Committee on Science, Space and Technology under the chairmanship of the Hon. George E. Brown, and as White House special assistant for science and technology public affairs.

Massimiano Bucchi is Professor of Sociology of Science at the University of Trento, Italy. He is a member of the PCST International Scientific Committee and has served as advisor and evaluator for several research and policy bodies, including the US National Science Foundation and the European Commission. His research addresses mainly the interaction among experts, citizens and policy-makers and the role of the public sphere in scientific debates. He is author of several books including *Science and the Media: Alternative Routes in Science Communication* (Routledge, 1998) and *Science in Society. An Introduction to Social Studies of Science* (Routledge, 2004), and co-editor of *Journalism, Science and Society: Science Communication between News* *and Public Relations* (with M. Bauer, Routledge, 2007) and essays in international journals including *Nature*, *Science*, and *Public Understanding of Science*.

Angela Cassidy is Research Fellow at the University of Leeds, UK, researching knowledge and communication of food chain risks among scientists, farmers, food campaigners, industry and the wider public in Britain. She has published a series of articles on popular evolutionary psychology, addressing the interactions between a newly emerging research subject: contemporary politics and the public domain. Her research interests centre on the relationship between expertise and everyday, commonsense knowledge and how this plays out in the communication, development and public legitimacy of the sciences.

Sharon Dunwoody is Evjue-Bascom Professor of Journalism and Mass Communication at the University of Wisconsin-Madison, USA. She studies components of the mediated science communication process, from the attitudes and behaviours of journalists and scientists who generate messages, to the efforts of audiences to process them. Co-edited books include (both with S. Friedman and C. Rogers): *Scientists and Journalists* (AAAS, 1986) and *Communicating Uncertainty* (Lawrence Erlbaum Associates, 1999). She has offered communication counsel on oversight committees of the National Academies and currently serves on the Committee on Public Understanding of Science and Technology of the American Association for the Advancement of Science.

Hester du Plessis is Senior Researcher in the Faculty of Art, Design and Architecture (FADA) at the University of Johannesburg, South Africa, and has tenure as Research Chair in Design Education at the National Institute of Design, Ahmedabad, India for 2008 and 2009. She co-authored a book with Gauhar Raza, *Science, Crafts and Knowledge* (Protea Book House, 2002). Her publications focus on linking traditional knowledge systems and the public understanding of science with traditional and modern design and technology. She heads a research niche area, 'Design for Development', in partnership with the Department of Industrial Design within FADA.

Edna Einsiedel is University Professor and Professor of Communication Studies in the Faculty of Communication and Culture at the University of Calgary, Alberta, Canada. Her research interests are in the social issues around life science technologies, specifically genomics and biotechnology, focusing on approaches to public engagement and participation and their institutional arrangements. She is co-leader on a GE3LS project (Genomics, Ethics, Economic, Environmental, Legal and Social Studies) on Genomics and Knowledge Translation in Health Systems, supported by Genome Canada. She has published in journals including *Science*, *Nature Biotechnology*, *Public Understanding of Science*, *Science Communication*, and *Science and Engineering Ethics*. She is currently editor of the international journal *Public Understanding of Science*. She is also a member of the Board of Governors for the Council of Canadian Academies of Science.

Iina Hellsten is Assistant Professor in the Faculty of Earth and Life Sciences, Free University Amsterdam, the Netherlands. Her main research interest has focused on the dynamics of metaphors in science communication. Her publications include *The Politics of Metaphor: Biotechnology and Biodiversity in the Media* (Tampere University Press, 2002) and she has published on public controversies regarding cloning, the Human Genome Project, genetically modified foods and stem cell research in *Science Communication, New Genetics and Society, Journal of Computer-Mediated Communication, New Media & Society, First Monday, Science as Culture, Scientometrics*, and *Metaphor and Symbol.* Her current research focuses on social responses to the bird flu threat.

Alan Irwin is Dean of Research and Professor at Copenhagen Business School, Denmark. His research deals with science and technology policy, scientific governance and science–public relations. His books include *Risk and the Control of Technology* (Manchester University Press, 1985), *Citizen Science* (Routledge, 1995), *Misunderstanding Science?* (with Brian Wynne, Cambridge University Press, 1996), *Sociology and the Environment* (Polity Press, 2001) and *Science, Social Theory and Public Knowledge* (with Mike Michael, Open University Press, 2003). He has published widely on European scientific governance, experiments in science and democracy, and public understandings of risk. He is a member of the 'Bioscience for Society' strategy panel of Britain's Biotechnology and Biological Sciences Research Council.

David A. Kirby was a practising evolutionary geneticist before leaving bench science to become Lecturer in Science Communication at the University of Manchester, UK. Several of his publications address the relationship between cinema, genetics and biotechnology, including essays in *New Literary History, Literature and Medicine*, and *Science Fiction Studies*. He is also exploring the collaboration between scientists and the entertainment industry, and has published in *Social Studies of Science* and *Public Understanding of Science* on this topic. He is working on a book entitled *Science on the Silver Screen: Science Consultants, Hollywood Films, and the Interactions Between Scientific and Entertainment Cultures.*

Robert A. Logan is a professor emeritus at the School of Journalism of the University of Missouri-Columbia, USA, where he was the Associate Dean for Undergraduate Studies and Director of the Science Journalism Center. He is on the senior staff of the US National Library of Medicine, where he assists in the evaluation and production of the Library's comprehensive health informatics services to the public. He has published more than 50 articles (including book chapters and refereed research about health campaigns), serves on the editorial boards of three journals, and is the first author of four books about science communication.

Federico Neresini teaches social research methodology and science, technology and society at the University of Padua, Italy. His main research interests are in the sociology of science, in particular public communication of science, social representations of science, and citizens' participation in decision-making processes about techno-scientific issues. He has focused more recently on biotechnology issues, with specific attention to *in vitro* fertilisation and cloning, and also on nanotechnology.

Brigitte Nerlich is Professor of Science, Language, and Society at the Institute for Science and Society, University of Nottingham, UK. She studied French, philosophy and linguistics in Germany and was previously a junior research fellow in general linguistics at the University of Oxford. Her current research focuses on the cultural and political contexts in which metaphors are used in public and scientific debates about genetics/genomics, infectious diseases and nanotechnology. She has written books and articles on the history of linguistics, semantic change, metaphor, metonymy, polysemy and, more recently, the social study of science and technology.

Giuseppe Pellegrini teaches social research methodology at the University of Padova, Italy. His current research focuses on social policy, citizenship rights and public participation, with specific regard to biotechnology issues. He is the coordinator of the research area 'Science and Citizens' at Observa – Science in Society. He is a member of the European Association for the Study of Science and Technology and of the Society for Social Studies of Science. His most recent book was *Biotecnologie e Cittadinanza* (*Biotechnology and Citizenship*) (Libreria Gregoriana Edizioni, 2005).

Hans Peter Peters is Senior Researcher at the Programme Group Humans-Environment-Technology of the Research Center Juelich, Germany, and Adjunct Professor for Science Journalism at the Free University of Berlin. His research focuses on how the public make sense of science, technology and the environment within a 'media society', and on the interactions between science and the media. He has coordinated an international team of scholars analysing biomedical researchers' contacts with the media in five countries. He is a member of the PCST International Scientific Committee.

Bernard Schiele is Head of the international PhD Programme in Museum Studies, researcher at the Interuniversity Research Centre on Science and Technology, and Professor of Communications at the Faculty of Communication at the University of Quebec at Montreal, Canada. He frequently teaches and lectures in North America, Europe and Asia. He has been working for a number of years on the socio-dissemination of science and technology. He is a member of several national and international committees and is a regular consultant on scientific culture to government bodies and public organisations. He is also a founding member and current member of the PCST International Scientific Committee. He chairs the International Scientific Advisory Committee for the New China Science and Technology Museum (opening 2009).

Brian Trench is Senior Lecturer and former Head of School in the School of Communications, Dublin City University, Ireland. He was a journalist for 20 years before joining the university, where he teaches science and society and science in the media, as well as study skills and research methods, and chairs the university's Research Ethics Committee. His research and publications have centred on science communication on the internet, models of public communication of science and technology, and discourses of the knowledge society. He was a member of the government's advisory body, Irish Council for Science Technology and Innovation, 1997–2003. He is a member of the PCST International Scientific Committee.

Jon Turney is a science writer and lecturer. He has taught at University College London, Birkbeck College and City University, UK, and currently leads the MSc in creative non-fiction at Imperial College, London, UK. He has reviewed popular science books widely in the press and in academic journals. His books include *Frankenstein's Footsteps: Science, Genetics and Popular Culture* (Yale, 1998). His current project is writing *The Rough Guide to the Future*, to be published in 2009.

Steven Yearley is Professor of the Sociology of Scientific Knowledge at the University of Edinburgh, UK, and also Director of the Economic and Social Research Council's Genomics Policy and Research Forum. He has a long-standing interest in environmental sociology and science studies, and has devised techniques for public engagement in environmental knowledge-making. Recently he has authored *Making Sense of Science* (Sage, 2005), *Cultures of Environmentalism* (Palgrave, 2005) and *The Sage Dictionary of Sociology* (with Steve Bruce, Sage, 2006).

Introduction

When Carl Sagan died in 1996, his had become a widely recognised name among those watching television in the USA and far beyond. An astronomer with a qualification also in biology, Sagan was an accomplished performer in public and on screen. His 22 books sold more than 10 million copies worldwide and his television series, *Cosmos*, first broadcast on the American PBS network in 1980, was seen by over 500 million people. He wrote more than 400 popular science articles for magazines.

Carl Sagan had the good fortune to be an astronomer at a time when space exploration began. As the cover illustration of this book indicates, he also had the good fortune to have, as the obituarist in *The Australian* newspaper (30 December 1996) put it, 'a compelling presence [and] good looks'. He also had 'an enviable talent to explain'.

Sagan represented some kind of high-water mark in the evolution of the public, visible scientist – and was perhaps an early example of a 'celebrity scientist'. Popularisers and explainers of science have a longer history than professional scientists – communicating ideas or insights drawn from scientific research to a wider public was part of the enlightenment enterprise of the 18th century; the term 'scientist' was first used in the mid-19th century.

As optimism grew about what science could bring, so the efforts to popularise science multiplied. Associations for the advancement of science, or similar bodies, were formed in many industrialised countries in the 19th century. There were more waves of enthusiasm for science, and for its popularisation, in the 20th century. In the years before, during and after World War II, a generation of scientists with twin commitments to political activism and to science sought to explain contemporary science and its benefits to the 'common man'. When the USA's confidence in its scientific, technological and political leadership was shaken by the launch of the Soviet sputnik, major initiatives started in the USA to spread the good news about science through schools and the mass media.

However, the science communication movement that developed in various parts of the world, particularly from the 1970s onwards, involved broader groups within the scientific communities and reflected, at least in part, a concern that the good news was not being heard – even that the respect and reputation of science needed to be salvaged. Very many of the active practitioners of science communication came from a background in the natural sciences, as did those who offered instruction for and analysis of these communication endeavours. In many cases they had acquired additional knowledge and skills, either through immersion in practice or – less frequently – through formal study.

The scientific communities' perspectives on the issues in science communication were also reflected in the developing field of science journalism, a specialism among professional journalists that grew in strength and presence alongside the broader field of science communication. Then, and to a large extent now, a background in the natural sciences was practically an entry requirement.

With the gradual professionalisation of science communication, courses have proliferated in universities or as part of professional development; until recently, most have been accommodated in science faculties or professional societies, targeted at science students, graduates or professionals, and often delivered by 'converted' scientists.

At the start of the 2000s, however, interest in science communication has broadened, and the influence of its origins in the natural sciences has weakened. This has been both cause and effect of the growing maturity and reflexivity of science communication. In its early defensiveness, science communication sought to define itself as distinct and different. Building its own networking and publishing structures has been a necessary part of growing up. But it has also, for some time, tended to obscure from view the interests shared with those in such longer-established disciplines as history of science, philosophy of science, science education and sociology of science, on the one side, and mass communication, journalism and cultural studies, on the other.

As science communication has matured, it has found it possible to recognise these commonalities and connections. Its area of operation has also grown, necessarily bringing contact with neighbours and relatives. Its appeal and relevance have attracted an increasing number of contributions from those of different backgrounds. All these developments have strengthened rather than weakened science communication, both as professional practice and as academic discipline.

Thus, unlike many existing and undoubtedly valuable manuals, this handbook does not aim to provide instructions or direct guidance on how to communicate science and technology; rather, it tries to offer a state-of-the-art map of a field that has developed substantially, not only at the level of practice, but also in terms of research and reflection as well in terms of the diversity and richness of points of view.

The contributors include mass communication scholars, sociologists, discourse analysts, public relations practitioners, journalists and journalism educators, and more. Several are established opinion-formers and frequently referenced authorities in the merging of science communication and related disciplines.

A key discussion ground for many of the contributors has been the international Public Communication of Science and Technology (PCST) network, which organises a biennial conference bringing together practitioners, educators and analysts in science communication, with backgrounds in natural science, social sciences and humanities, or in no particular academic discipline. The editors of this volume and several of the contributors are members of the PCST International Scientific Committee, which shapes the conferences.

The book has benefited from, and is intended as a contribution to, the discussions at these and similar forums, as well as in the many and varied science communication courses, programmes, projects and departments in the wider PCST field internationally. We aim to support formal study in the topics addressed, and to provide insights and promote reflection among those professionally engaged in science communication initiatives.

The organising categories for commissioning the contributions to this volume were *actors* (who is involved in science communication?); *arenas* (what means do they use? in what spaces do they operate?); and *issues* (what challenges and opportunities do they face?). As we might have anticipated, the boundaries between these categories have broken down, and almost all contributions address a mixture of these categories. The more important connections and distinctions have emerged around questions of the underlying assumptions in PCST, the shifts in approach over time, and the increasing diversity of media and mediators. From different starting points, several contributors have focused on the complexities of communicating uncertainty and risk, or on the ways in which changing (or apparently changing) strategies in PCST are linked to broader societal transformations.

Several chapters – and the structure of this handbook – try also to reflect the gradual shift in policy discourse from keywords such as 'popularisation' and 'public understanding of science' to 'dialogue', 'engagement' and 'participation'. In this light, the importance of locating science communication in the wider context of science in society is also emphasised in this collection.

Inevitably, there are gaps in the coverage and questions raised that need further elaboration. Science communication takes place in many more spaces than are addressed in detail here: the role of the broadcast media, for example, merits much more analysis. There is far more variation between public cultures and scientific cultures than is reflected in this volume; we are acutely aware of an anglophone bias in the countries referred to and examples used. There is more to be learned than we have been able to include from cultural or education studies about how publics make sense of science.

We hope, nevertheless, that this volume will contribute to greater stability and clarity of conceptualisation and strategising in science communication as well as to the continuing maturing of practice and theory in this field.

At the end of each chapter, books and articles are suggested for the reader wishing to learn more about the topics addressed, and beyond. We hope this handbook will also be used as a jumping-off point for further reflection and, undoubtedly, contestation.

The editors wish to acknowledge the contribution of Bruce Lewenstein of Cornell University in initiating this book project. We also acknowledge the support of the European Commission and the Istituto Veneto di Scienze, Lettere ed Arti, in hosting a seminar in Venice in January 2007 at which many of the contributors were able to discuss preliminary versions of their own chapters with their colleagues.

Massimiano Bucchi, Trento
Brian Trench, Dublin
March 2008

1

Popular science books

Jon Turney

Introduction

Visit any sizable city bookstore these days, and there will be a shelf labelled 'popular science'. This is an established publishing category, and there has been something of a boom in such books over the past 25 years or so. This is intriguing, for a number of reasons. Books are only one of a number of ways in which the contemporary media package representations of science. But they are extremely versatile, relatively cheap to produce, and are suited to extended exposition in ways that other, more modern media find hard to emulate. Books also have a cultural cachet that attracts writers – and readers – and can lend them influence beyond their immediate audience. The printed book also pre-dates the rise of modern science, so contemporary cultural products invite comparisons with earlier efforts to communicate science. But such comparisons must be approached with caution. The organisation of science, and of science communication, is very different in the modern era, and continues to change, and the same applies to the production and consumption of media. However, the constancy of the book, in one role or another, may illuminate important features of the relationship between science and public that are more difficult to explore through other media. This chapter offers a brief reflection on the idea of 'popular science' – a more problematic term than its use by booksellers and publishers may suggest. Then I offer some comments on the history of the book as a vehicle for disseminating scientific ideas, and finally I expand on some of the ways in which science books have been studied, or could be studied.

Popular science: popular? scientific?

The term 'popular science' usually implies an attempt, in some medium, to make scientific ideas accessible to non-professional audiences. Myers (2003) emphasises that 'popular science' is an unusual category because its scope is defined by what it is not. On the one hand, it is not technical science, as would appear in a journal. On the other, as distinct from 'pop science' (Basalla 1976), it relates to non-fictional representations of science. This is not the place for an extended discussion of what constitutes non-fiction, but let us say that in this context it denotes a reference to what are claimed to be actual, real events or persons, and genuinely scientific ideas. Any of these things is open to dispute, and there are plenty of books that deliberately blur the already fuzzy boundaries between fiction and non-fiction. These include adult novels such as those about cloning by David Rorvik (1978) or about climate change by Michael Crichton (2004), which are deliberate interventions in current controversy; and children's books such as the *Uncle Albert* series by British physicist Russell Stannard (e.g. Stannard 1989), which use storytelling as a vehicle for conveying scientific concepts. However, these are far outnumbered by examples where the intent to convey non-fiction truths is carried through more straightforwardly, and these are worth studying in their own right. Although there is scope for much instructive discussion of exactly how to define the category, and whether it has the characteristics of a genre (Jurdant 1993), publishers and bookstores do seem to agree on what goes onto a shelf labelled 'popular science'. Today, the titles to be found there are typically written by professional scientists or science writers. Their styles, motives and approaches may vary widely, but at least part of their aim is to communicate some of what they take to be science.

We must still ask, though: are such works either scientific or popular? They certainly need not be popular in the sense of commanding a large audience. The term again refers to intent – they could be accessed by many people, even if they only turn out to appeal to a minority. This raises difficult questions about what is usually called the 'level', where higher level means more technical, closer to the actual science – in some way – and lower level is less so. One difficulty for writers is always that the ideal, if such it is, of assuming no prior knowledge conflicts with the necessity to get on and deal with the exposition one actually wants to present. Another metaphorical approach may be to consider the degree of dilution of the science, although there is no way of measuring this.

This leads to a caution about a well-known critique of the 'dominant' view of science popularisation (Hilgartner 1990). That critique takes issue with the view that popularisation is a separate activity from professional science communication, and that scientists are responsible for the canonical versions of what counts as knowledge, which others can only approach more or less closely, but must always defer to. It has also been developed by Whitley (1985) and Cooter and Pumfrey (1994; see also Bucchi, Chapter 5 in this volume).

It is true that much is obscured by such a one-dimensional view of popular science, and that it is more informative to see popular versions as a large ensemble of appropriations or re-creations of science, with different motives and realised in different media. The best studies cited below usually adopt this assumption. But there is still sense in considering whether a work avowedly intended for popular consumption is more or less technical in content. Crudely, it makes a difference if you offer an account of modern physics with all the equations left in, as Roger Penrose (2004) has done in *The Road to Reality*, or write a 'biography' of an equation containing no mathematics at all aside from the four symbols in the title, as the award-winning writer David Bodanis (2000) did in $E = MC^2$. As the author of a review that described the latter book as presenting science in homeopathic dilution, I think there is a spectrum here on which popular works can be meaningfully placed.

Popular science books: a historical perspective

Books are an attractive focus for a study of popularisation, partly because of their long-established significance in cultural production. The technology of the printed book has been around in the west since the 15th century, so its history encompasses of modern science. Anyone with ambitions to consider popularisation in the round has to take account of an increasing diversity of media in recent times – and books must often be considered in relation to such media, as we shall see. I emphasise again that the continuity of the book must not obscure changes in reading publics, markets and publishing practice, or the sweeping changes in science itself (Knight 2006). Nevertheless, the book form is a notable constant in the history of science communication, and studying that history is an important job in its own right.

The history of professional science communication is the story of the rise of the specialist journal, with its formal style and technical language. Such formality is efficient for addressing peers, but excludes everyone else. The separation between science and public that this system embodies is what recent 'popularisers' claim to try and overcome.

So one might imagine that earlier instances of such a separation would give rise to efforts at what we would now label popularisation. A notable example is Newton's *Principia*. Not because it was written in Latin, which was accessible to the educated gentry who were the bulk of the literate reading public of the late 17th century, but because it was mathematically impenetrable. As the cult of Newton grew in the following century, many popular lectures and books, including a celebrated account of his ideas by Voltaire, set out versions of the Newtonian system that were more accessible. There were even children's versions, such as the pseudonymous effort by Tom Telescope of 1861, attributed to the pioneering children's author John Newbery but possibly written by Oliver Goldsmith or Christopher Smart (Secord 1985). The full title of this notably successful work – *The Newtonian System of Philosophy, explained by familiar objects, in an entertaining manner, for the use of young persons* – expresses aims that many later popular writers would also embrace.

Such works grew in number and variety as disciplines developed and divided, more science became susceptible to mathematical formalisation, and science in general was professionalised. Our view of this history is easily distorted by the prominence of particular books or authors. Much discussion centres on key works such as *The Origin of Species* in 1859, which expounded pathbreaking science, but were still accessible to a generally educated readership. Or it tends to focus on periods when well-known scientists wrote for popular audiences – as in the 1930s when Arthur Eddington and James Jeans published best-selling versions of Einsteinian physics and astrophysics.

However, closer historical study generally reveals that less-celebrated writers, their books now forgotten, were active as expositors of science throughout the past two centuries. By Thomas Huxley's time in Victorian England, it was possible to support oneself (precariously) as a writer specialising in natural history and science. But, as Fyfe (2005) puts it, 'by the 1850s an increasingly clear division had developed between writings that contributed to a scientific reputation (and usually paid little) and those that paid the bills (but did nothing for one's reputation)'. Her study, set against the backdrop of the shift from the public sphere of the 18th century to the mass audience of the 19th, shows how a market emerged for a whole range of science books, aimed at instruction, moral improvement or even entertainment.

Increasingly, authors specialised in science or its exposition, and many authors of science books for lay readers were not active as scientific researchers. Among historians, as Bowler (2006) has argued, there arose a widespread assumption that the vast majority of scientists turned their back on writing for lay readers in the early 20th century as professionalisation took hold. But, he suggests, this was not really true. There were some scientists who were happy to write for the wider public, and whom publishers valued because their credentials allowed them to sell the books as 'educational'. But there was also a mixed economy, with a number of now little-known authors, usually without professional scientific training, writing more 'entertaining' works.

It is perhaps true that few prominent scientists wrote for the public in this period, but by the 1930s a cadre of 'visible scientists' (Goodell 1977) was again active. Retrospective reviews generally lump together the conservative British physicists Eddington and Jeans, on the one hand, and the more liberal or radical biologists, such as Julian Huxley and J. B. S. Haldane, on the other. But there were also notable popular works by Einstein, for example, whose global fame boosted the appeal of a number of expositions of special and general relativity that remain in print today.

After World War II, the picture shifted again – there was not only a large literate public, but also a larger cohort with higher education. Again, accounts of the general development of popular science books differ according to the focus of the author. Lewenstein (2005) detects a major shift in the prominence of science books in the 1970s, but his verdict draws from studies of Pulitzer Prize winners and best-sellers listed by the *New York Times* in this period. As Lewenstein puts it: 'beginning with Carl Sagan's *Dragons of Eden* in 1978, then every year or every other year the Pulitzers begin honoring a science book ... Clearly something happens in the late 1970s to make science books more central to American culture. Science becomes a part of the general public discussion' (Lewenstein 2005).

The point is well made, and the same shift can be detected in Great Britain (where science books have now had their own prize for 20 years). The influence of other media was powerful, with Sagan's TV series *Cosmos* leading to a book that became a best-seller on both sides of the Atlantic (Sagan 1980). The interpretation of the cosmic story that unfolds from the Big Bang, the integrated grand narrative that Eger (1993) has dubbed 'the new epic of science', was argued over by Steven Weinberg, Sagan, Steven Hawking, and many other authors (Turney 2001). The modern popular science book boom was finally under way.

Again, this broad-brush story is too simple – there were also more gradual changes that were visible earlier. In the USA, the advent of mass higher education brought a new reading public to non-fiction. A close study of Knopf best-sellers, now largely forgotten trade books about human history and evolution in the 1950s and 1960s, shows how they came to depend less on old models. In the early post-war period, this publisher relied on a biographical style closely akin to that used by researcher-turned-writer Paul deKruif (1926) in his successful *Microbe Hunters*. But as the years went by, they found that books with considerably more technical content could be promoted successfully and sold well. The overall story of the subject, incorporating up-to-date results and told by an expert or a journalist in close touch with the experts, proved as appealing as the stories and struggles of the discoverers (Luey 1999).

Both models continued to exist through the popular science boom – now declared by some to be over (Tallack 2004) – and beyond. So we now have a profusion of titles in print, on a wide range of subjects, offering many different treatments of similar subjects. Some are widely regarded as classics, and remain strong sellers over many years. The leading example is perhaps *The Selfish Gene* (Dawkins 1976), which was expanded into a successful second edition in 1989, and in 2006 republished in a 30th anniversary edition along with a companion volume of essays considering its influence. This kind of prominence is exceptional, but testifies to the influence a science book may have. With the number of different titles in print so high, their collective influence must be greater still. What are we to make of it?

Ways to read a science book

This brief sketch of the history of popular science books establishes, at least, that they are numerous and diverse. Books are complex objects, and can be studied in many different ways, most of which have been applied to popular science at one time or another. The following examples are restricted to studies published in English, and mainly to studies of English language works, because of the author's linguistic limitations, not because popular books in other languages are not important or equally worthy of scholarly attention.

Aside from the basic issues of what books existed when, what was in them, and who wrote them – not always easy to establish for popular books in earlier periods – other historical approaches can yield rich dividends. The most difficult to realise is probably to try and recover a sense of how a book was read, as Secord (2000) does brilliantly for Robert Chambers' early 'epic of science', *Vestiges of the Natural History of Creation*. The core of Secord's study is an analysis of readers' responses to *Vestiges*' natural-historical account of the evolution of the universe, culled from diaries, letters, reviews and newspaper and journal commentaries. It is no disrespect to more strictly literary studies to observe that this kind of reconstruction requires immense enterprise on the part of the investigator, which is perhaps why this work remains unmatched. A recent book-length study of Rachel Carson's (1962) *Silent Spring* (Murphy 2005) offers an interesting comparison in the modern era, concerned with the way a book can influence a political debate. In this case, the fact that Carson was already a nationally celebrated author, that *Silent Spring* was serialised in *The New Yorker* before publication, and that it inspired a CBS *60 Minutes* TV programme were all key factors in increasing the book's impact (Kroll 2001).

Authors like Carson, whose *The Sea Around Us* (Carson 1951) was a notable best-seller long before *Silent Spring*, also attract biographers, and the volumes devoted to her (Lear 1997) and to Sagan (Davidson 1999) contain much useful material about the context of their work. *Silent Spring* is also the subject of a collection of essays devoted to Carson's rhetoric (Waddell 2000), the most interesting contribution to which is an analysis of its deployment of cold war rhetoric (Glotfelly 2000).

Sagan, too, has been the subject of close rhetorical analysis (Lessl 1985, 1989), and this is perhaps the most common mode of study of recent popular science writing.

More straightforwardly, literary analyses are also fairly numerous. A fair number consider individual works, or a single author. Not surprisingly, these tend to be the better known authors such as Steven Jay Gould (Fuller 1990) or James Gleick (Kelley 1993; Paul 2004). Some other important examples are collected by McRae (1993).

The volume and diversity of recent popular science make it difficult for authors to range more widely. Can these works be classified into sub-categories? One useful attempt (Mellor 2003) suggests that there are three main types. 'Narratival' works relate an episode in the history of science or the life of a single scientist. 'Expository' books are mainly concerned with the exposition of a particular discipline. And 'investigative' works are more journalistic in style and take up topical or controversial subjects. Whether these categories are exhaustive or mutually exclusive is open to discussion, but Mellor's further suggestion that it is expository works that tend to be labelled as popularisations by authors, publishers and booksellers seems accurate.

Focusing on just one of these categories makes broader study more feasible, as in the first book-length study of recent popular physics (Leane 2007). This landmark effort is mainly literary in approach, but reads recent works in the context of the history of popularisation, the 'two cultures' debate and the 'science wars' of the 1990s. The overall assumption is that 'popular science . . . mediates between science and literature by presenting the content of the former through some of the established techniques of the latter'. Those techniques include figurative language, narrative and characterisation, and all receive attention in commentaries on a range of notable recent popular books in the physical sciences. Especially novel is the reading of Gleick's (1987) *Chaos* and Waldrop's (1992) *Complexity* as adopting character stereotypes from the detective novel.

Aside from literary approaches, there are significant studies in this area from publishing history, notably an important investigation of Eddington's and Jeans' work in the 1920s and 1930s (Whitworth 1996). But most other published work falls into a category that can loosely be labelled 'science studies'. This is not clearly demarcated from literary analysis, any more than popularisation is simply divided from professional science communication. But research or critique that makes use of ideas from philosophy, sociology or political studies of science can usefully be considered under this heading.

Such studies include, for example, a content analysis of Gould's writing, popular and professional, in terms of his position on the history and philosophy of science (Shermer 2002). Also philosophical in intent is Curtis's (1994) important analysis of 'Baconian' narratives in popular science as encoding an implicit epistemology. Philosophers have also turned their attention to popular science on occasion, usually to be highly critical, whether in the 1930s (Stebbing 1937) or in the 1990s (Midgley 1992).

Also in the realm of critique, Mellor (2003), already cited as the source of a useful categorisation, argues that popular physics books are, by and large, constructed to circulate images of science that reflect scientists' interests and counter other images of science that may arise in other media. Hedgecoe (2000) suggests that the popularisation of genetics at that time contributed to the overreaching claims for genetic explanation, which have been labelled geneticisation.

Future work will doubtless develop both literary approaches and science studies critiques, enriched by linguistic analyses such as those offered by Myers (1990, 1994). It may also develop in several other directions. One is to probe in more detail what may be special about popular science books as expository texts, perhaps in terms of how they bring off persuasive explanations of complex or highly mathematical material (Turney 2004). There is also enormous scope for studies that integrate work on books with examination of the role of other forms of science communication, such as the web of text that Fred Hoyle wove over many years to promote his cosmological theories and his ideas about the origins of life (Gregory 2003). And modern children's science books are a growing area of publishing that is ripe for study, now just beginning (Bell 2007a, 2007b).

Finally, we may ask whether the study of popularisation will affect its practice. Some disciplines within science studies are becoming more conscious of ways in which the possibilities of popularisation relate to their own position in the public sphere. This may be a matter of professionals trying to produce popularisations of key ideas in their area, as has been done successfully for the sociology of scientific knowledge (Collins and Pinch 1993). More problematic are expressions of frustration on the part of professional historians of science about the way stories from history are rendered by popular authors (Miller 2002; Charney 2003). The challenge is to do better.

And the challenge is considerable, because the contextual approach emphasised by the majority of practising historians is a poor fit with the biographical conventions of popular genres. It is interesting that the only recent one-volume history of science addressed to a popular audience is by a well-known science writer, John Gribbin (2003). However, there are also signs that some younger science writers are alive to the possibilities of histories that take account of more up-to-date approaches in the discipline – books such as Oliver Morton's (2002) *Mapping Mars* and Carl Zimmer's (2004) *Soul Made Flesh* offer a skillful synthesis of storytelling and social history, which suggests that the possibilities of popular science books are far from exhausted.

As for the study of such books, one may hope for more work that exploits the possibilities for close comparison between 'specialised' science texts and popularisations of the same subject, or even by the same author. This approach, exemplified in different ways by Fahnestock (1986); Myers (1990); Schickore (2001), lends itself more easily to studies of shorter texts – a paper and a popular article, say. But it also offers much scope for richer analyses of popular books. Schickore's general view, informing her satisfyingly detailed study of 19th century German physicist and physiologist Hermann von Helmholtz as popular writer and scientist, is that a popular text can be a 'complex, multilayered piece of writing, directed to several audiences, manifesting a delicate balance between conflicting agendas'. This is doubtless correct, if unsurprising. But her further recommendation about method is one that future students in this area will do well to keep in mind: 'Studies of science popularization should pay attention to ambiguities and tensions within the popular writings themselves. These tensions serve as indicators of various different agendas underlying the production of the text.'

Suggested further reading

Charney, D. (2003) 'Lone geniuses in popular science: the devaluation of scientific consensus', *Written Communication*, 20: 215–41.

Leane, E (2007) *Reading Popular Physics – Disciplinary Skirmishes and Textual Strategies*, Aldershot: Ashgate.

McRae, M. W. (ed.) (1993) *The Literature of Science: Perspectives on Popular Scientific Writing.* Athens, GA: University of Georgia Press.

Shinn, T. and Whitley, R. (eds) (1985) 'Expository science: forms and functions of popularisation', *Sociology of the Sciences Yearbook 9*, Dordrecht: Reidel.

Turney, J. (2004) 'Accounting for explanation in popular science texts: an analysis of popularized accounts of superstring theory', *Public Understanding of Science*, 13: 331–46.

Other references

Basalla, G. (1976) 'Pop science: the depiction of science in popular culture', in Holton, G. and Blanpied, W. A. (eds), *Science and its Public: The Changing Relationship*, Boston Studies in the Philosophy of Science 33, Dordrecht: Reidel, 260–78.

Bell, A. (2007a) 'Chasing the Cheshire Cat: "invisible" boundaries of credibility in children's science literature', in Ramone, J. and Twitchen, G. (eds), *Boundaries*, Newcastle: Cambridge Scholar's Publishing.

—— (2007b) 'What Albert did next: the Kuhnian child in science writing for young people', in Pinsent, P. (ed.) *Time Everlasting: Representations of Past, Present and Future in Children's Literature*, Lichfield: Pied Piper Publishing.

Bodanis, D. (2000) *$E = MC^2$: A Biography of the World's Most Famous Equation*, London: Macmillan.

Bowler, P. J. (2006) 'Experts and publishers: writing popular science in early twentieth-century Britain, writing popular history of science now', *British Journal for the History of Science*, 39: 159–87.

Carson, R. (1951) *The Sea Around Us*, Oxford: Oxford University Press.

—— (1962) *Silent Spring*, New York: Houghton Mifflin.

Charney, D. (2003) 'Lone geniuses in popular science: the devaluation of scientific consensus', *Written Communication*, 20: 215–41.

Collins, H. and Pinch, T. (1993) *The Golem – What You Should Know About Science*, 2nd edn, Cambridge: Cambridge University Press.

Cooter, R. and Pumfrey, S. (1994) 'Separate spheres and public places: reflections on the history of science popularization and science in popular culture', *History of Science*, 32: 237–67.

Crichton, M. (2004) *State of Fear*, New York: Harper Collins.

Curtis, R. (1994) 'Narrative form and normative force – Baconian storytelling in popular science', *Social Studies of Science*, 24: 419–61.

Davidson, K. (1999) *Carl Sagan – A Life*, Chichester: John Wiley.

Dawkins, R. (1976) *The Selfish Gene*, Oxford: Oxford University Press.

DeKruif, P. (1926) *Microbe Hunters*, New York/London: Harcourt Brace Jovanovich.

Eger, M. (1993) 'Hermeneutics and the new epic of science', in McRae, M. W. (ed.) *The Literature of Science: Perspectives on Popular Scientific Writing*, Athens, GA: University of Georgia Press, 186–209.

Fahnestock, J. (1986) 'Accommodating science: the rhetorical life of scientific facts', *Written Communication* 30: 275–95, reprinted in McRae, M. W. (ed.) *The Literature of Science: Perspectives on Popular Scientific Writing*, Athens, GA: University of Georgia Press.

Fuller, G. (1998) 'Cultivating science: negotiating discourse in the popular texts of Stephen Jay Gould', in Martin, J. R. and Veel, R. (eds), *Reading Science: Critical and Functional Perspectives on Discourses of Science*, London: Routledge, 35–62.

Fyfe, A. (2005) 'Conscientious workmen or booksellers' hacks? The professional identities of science writers in the mid-nineteenth century', *Isis*, 96: 192–223.

Gleick, J. (1987) *Chaos: Making a New Science*, New York: Viking Penguin.

Glotfelly, C. (2000) '*Silent Spring*: the trope of war in modern environmentalism', in Waddell, C. (ed.) *And No Birds Sing: Rhetorical Analyses of Rachel Carson's* Silent Spring, Carbondale, IL: Southern Illinois University Press, 157–73.

Goodell, R. (1977) *The Visible Scientists*, Boston, MA: Little, Brown.

Gregory, J. (2003) 'Popularization and excommunication of Fred Hoyle's "life-from-space" theory', *Public Understanding of Science*, 12: 25–46.

Gribbin, J. (2003) *Science: A History, 1534–2001*, London: Allen Lane.

Hedgecoe, A. (2000) 'The popularization of genetics as geneticization', *Public Understanding of Science*, 9: 183–9.

Hilgartner, S. (1990) 'The dominant view of popularisation: conceptual problems, political uses', *Social Studies of Science*, 20: 519–39.

Jurdant, B. (1993) 'Popularization of science as the autobiography of science', *Public Understanding of Science*, 2: 365–73.

Kelley, R. T. (1993) 'Chaos out of order: the writerly discourse of semipopular scientific texts', in McRae, M. W. (ed.) *The Literature of Science: Perspectives on Popular Scientific Writing*, Athens, GA: University of Georgia Press, 132–51.

Knight, D. (2006) *Public Understanding of Science: A History of Communicating Scientific Ideas*, London: Routledge.

Kroll, G. (2001) 'The "Silent Springs" of Rachel Carson: mass media and the origins of modern environmentalism', *Public Understanding of Science*, 10: 403–20.

Leane, E. (2007) *Reading Popular Physics – Disciplinary Skirmishes and Textual Strategies*, Aldershot: Ashgate.

Lear, L. (1997) *Rachel Carson: Witness for Nature*, New York: Henry Holt.

Lessl, T. M. (1985) 'Science and the sacred cosmos: the ideological rhetoric of Carl Sagan', *Quarterly Journal of Speech*, 71: 175–87.

—— (1989) 'The priestly voice', *Quarterly Journal of Speech*, 75: 183–97.

Lewenstein, B. V. (2005) 'Science books since World War II', unpublished (in press for volume five of *The History of the Book in America*, http://people.cornell.edu/pages/bvl1/#Publications).

Luey, B. (1999) 'Leading the public gently: popular science books in the 1950s', *Book History*, 2: 218–53.

McRae, M. W. (ed.) (1993) *The Literature of Science: Perspectives on Popular Scientific Writing*, Athens, GA: University of Georgia Press.

Mellor, F. (2003) 'Between fact and fiction: demarcating science from non-science in popular physics books', *Social Studies of Science*, 33: 509–38.

Midgley, M. (1992) *Science as Salvation: A Modern Myth and its Meaning*, London: Routledge.

Miller, D. P. (2002) 'The "Sobel Effect": the amazing tale of how multitudes of popular writers pinched all the best stories in the history of science and became rich and famous while historians languished in accustomed poverty and obscurity, and how this transformed the world. A reflection on a publishing phenomenon', *Metascience*, 11: 185–200.

Morton, O. (2002) *Mapping Mars: Science, Imagination and the Birth of a World*, New York: Fourth Estate.

Murphy, P. (2005) *What A Book Can Do: The Publication and Reception of Silent Spring*, Amherst, MA: University of Massachusetts Press.

Myers, G. (1990) 'Making a discovery: narratives of split genes', in Nash, C. (ed.) *Narrative in Culture – The Uses of Storytelling in the Sciences, Philosophy and Literature*, London: Routledge, 103–26.

—— (1994) 'Narratives of science and nature in popularizing molecular genetics', in Coulthard, M. (ed.) *Advances in Written Text Analysis*, London: Routledge, 179–91.

—— (2003) 'Discourse studies of popular science: questioning the boundaries', *Discourse Studies*, 5: 265–79.

Paul, D. (2004) 'Spreading chaos: the role of popularizations in the diffusion of scientific ideas', *Written Communication*, 21: 32–68.

Penrose, R. (2004) *The Road to Reality: A Complete Guide to the Laws of the Universe*, London: Jonathan Cape.

Rorvik, D. (1978) *In His Image: The Cloning of a Man*, London: Hamish Hamilton.

Sagan, C. (1978) *The Dragons of Eden: Speculations on the Evolution of Human Intelligence*, New York: Random House.

Sagan, C. (1980) *Cosmos*, New York: Random House.

Schickore, J. (2001) 'The task of explaining sight – Helmholtz's writings on vision as a test case for models of science popularization', *Science in Context*, 14: 397–417.

Secord, J. (1985) 'Newton in the nursery: Tom Telescope and the philosophy of tops and balls, 1761–1838', *History of Science*, 23: 127–51.

—— (2000) *Victorian Sensation: The Extraordinary Publication, Reception, and Secret Authorship of Vestiges of the Natural History of Creation*, Chicago, IL: University of Chicago Press.

Shermer, M. B. (2002) 'This view of science: Stephen Jay Gould as historian of science and scientific historian, popular scientist and scientific popularizer', *Social Studies of Science*, 32: 489–524.

Shinn, T. and Whitley, R. (eds) (1985) *Expository Science: Forms and Functions of Popularisation, Sociology of the Sciences Yearbook 9*, Dordrecht: Reidel.

Stannard, R. (1989) *The Time and Space of Uncle Albert*, London: Faber.

Stebbing, L. S. (1937) *Philosophy and the Physicists*, London: Methuen.

Tallack, P. (2004) 'Echo of the Big Bang: an end to the boom in popular science books may actually raise standards', *Nature*, 432: 803–4.

Turney, J. (2001) 'Telling the facts of life: cosmology and the epic of evolution', *Science as Culture*, 10: 225–47.

—— (2004) Accounting for explanation in popular science texts: an analysis of popularized accounts of superstring theory, *Public Understanding of Science*, 13: 331–46.

Waddell, C. (ed.) (2000) *And No Birds Sing: Rhetorical Analyses of Rachel Carson's* Silent Spring, Carbondale, IL: Southern Illinois University Press.

Waldrop, M. (1992) *Complexity: The Emerging Science at the Edge of Order and Chaos*, New York: Simon and Schuster.

Whitley, R. (1985) 'Knowledge producers and knowledge acquirers: popularisation as a relation between scientific fields and their publics', in Shinn, T. and Whitley, R. (eds) (1985) *Expository Science: Forms and Functions of Popularisation, Sociology of the Sciences Yearbook 9,* Dordrecht: Reidel, 3–37.

Whitworth, M. (1996) 'The clothbound universe: popular physics books, 1919–1939', *Publishing History*, 40: 53–82.

Zimmer, C. (2004) *Soul Made Flesh*, Oxford: Heinemann.

2

Science journalism

Sharon Dunwoody

Introduction

Most citizens who have completed their formal education continue to learn about science chiefly from mediated channels, those ubiquitous packagers of information for large numbers of readers, listeners and viewers. In most cases, these citizens encounter science information almost inadvertently, as they watch TV news, read their morning newspaper, or page through a magazine from the corner newsstand. A subset of the public goes out of its way to seek this material, creating opportunities for niche publications. One channel, the World Wide Web, is increasingly the channel of choice for both science information grazers and seekers.

This chapter offers a brief look at the science content of these important information channels, as well as at the specialised communicators who provide that content. It first tracks the historical evolution of science journalism, with particular reference to the USA, then moves to characteristics of today's science journalists, science journalism and the media outlets in which they are present, as described and analysed in published studies. The chapter ends by describing a few trends that are already changing the science journalism landscape.

A bit of history

Science stories have appeared in the mass media for as long as these channels have existed. Who wrote those stories, on the other hand, has varied over time and, probably, across cultures. Permit me to illustrate this evolutionary point with the US experience. Historian John Burnham wrote a seminal book in the 1980s chronicling the popularisation of science in the USA in the 19th and 20th centuries (Burnham 1987); I will capture a bit of that vivid history here.

By the late 19th century, several popular science magazines were already established in the USA – pre-eminent among them *Scientific American* and *Popular Science*

Monthly – and newspaper editors, keen to fill their pages, were happy to reprint texts of science lectures and to publish scientists' reflections on natural phenomena such as meteor showers. Notable in this era was the determination with which scientists participated in public communication venues. Scientists in the latter part of the 19th century viewed popularisation as part of their job *as scientists*. They felt they had useful knowledge to impart, sensed a need for public support, and readily employed the mediated channels of the day to share their stories of discovery. For example, at one point in the late 1800s all the officers of the then-dominant scientific society in the USA, the American Association for the Advancement of Science, were not only distinguished researchers but also authors who had published in one or more of the popular science magazines of the day.

The beginning of the 20th century saw increasing specialisation in science, which left scientists little time to engage in popularisation. Exacerbating the trend was the growing professionalisation of science, which pushed scientists to see themselves as individuals more skilled than, and apart from, everyday people. As scientists developed their own languages, their own training regimens and their own reward systems, communication with 'others' became a low priority. To make matters worse, major scientific societies began to punish scientists for daring to popularise, by ostracising offending individuals and even denying them access to rewards such as membership in honorific societies. A case from the mid-1960s with which I had direct contact illustrates this situation: an American scientist studying Meniere's disease, a debilitating condition of the inner ear, was denied membership in an honorific society because he had been named in a newspaper story about his research; the society had deemed the story to be 'unethical advertising'. Goodell's classic book *The Visible Scientists* (Goodell 1977) is replete with examples of how even senior, accomplished scientists were subjected to sustained negative repercussions as a result of their popularising efforts. Although, as I argue at the end of this chapter, popularisation has again become *au courant* for many scientists, residual hostility within the scientific culture makes it a risky behaviour even today. But back in the early years of the 20th century, too much investment in popularisation could ruin a scientist's career, so scientists by and large left the world of popularisation to journalists and the mass media.

The mass media's interest in science remained steady throughout this period. The technology of warfare, discoveries of planets and entire galaxies (not to mention the Martian canals), and advances in medical care were easy for journalists to 'sell' to their editors. These editors did not care that a topic was scientific, only that it was novel and likely to grab the attention of their readers. Canvas the early issues of any newspaper and you are likely to find a litany of stories that we would classify today as 'science' in the broadest sense.

Still, few journalists in the USA by the mid-20th century would have defined themselves as science writers. Specialist reporters are expensive and, consequently, rare in most media organisations. Editors believed strongly in the ability of a good generalist to cover anything, and worried more about the by-products of cosy relationships between journalists and their sources than about the need to apply specialised knowledge to complex topics. Through much of the 20th century, a common practice in American news media was to rotate reporters across beats every few years to prevent the pitfalls of reporter/source intimacy.

A few specialised science reporters did gain a foothold in newspapers early in the 20th century, most of them at major newspapers and wire services. In the 1930s, this hardy band formed the National Association of Science Writers, in part to encourage American media organisations to value speciality reporting. The technological innovations catalysed by World War II, the Federal Government's post-war decision to invest in scientific research, the space race of the 1960s and the growing environmental concerns of the 1970s and 1980s galvanised many American media organisations, who scrambled to find science and environmental reporters to cover what looked like some of the major stories of the century. Despite the boom-and-bust nature of hiring patterns that characterise most mass media industries, science, environmental and health reporters now number in the thousands in the USA.

Although every culture has engendered its own catalytic chain, the results have been the same: numbers of science reporters have burgeoned in many countries – see, for example, Metcalfe and Gascoigne (1995) on changes in Australia. In addition to country-specific organisations of science writers, there are now global associations such as the World Federation of Science Journalists, and many students all over the world are expressing an interest in science communication training. Increased numbers of journalists typically lead to increased media coverage, and a number of longitudinal studies demonstrate such an accretion in science stories over time (Metcalfe and Gascoigne 1995; Bucchi and Mazzolini 2003).

That said, it is always important to remember that science reporters – like most classes of specialist reporter – remain a small subset of all reporters. Even in the best of times, most media organisations do without them. Thus science stories remain relatively minor components of media coverage. Dimopoulos and Koulaidis (2002), in an analysis of science coverage by four Greek newspapers, for example, found that the proportion of the news hole given over to science ranged from 1.5 to 2.5 per cent, similar to what Pellechia (1997) found in the USA and to what Metcalfe and Gascoigne (1995) found in Australia. In contrast, Dimopoulos and Koulaidis (2002) noted that political coverage accounted for some 25 per cent of stories, while sports made up 15 per cent in Greek newspapers.

The practice of science journalism today

The coverage of science is a topic of great interest to journalists, scientists and science communication scholars, occasioning much research and commentary. Here I examine a few key themes arising from those analyses and discussions.

Science news is overwhelmingly print-oriented and focused on biomedical topics

Although longitudinal studies of media content are expensive and therefore rare, the available studies indicate that science journalism has been, and remains, dominated by print media outlets: most journalists who define themselves as science specialists work for newspapers, magazines and newsletters; radio and television science journalists represent a much smaller proportion. The internet offers another outlet for these specialists, a topic I return to below.

For media outlets in many countries, the bulk of what passes for science writing is about medicine and health. Bauer (1998) tracked what he called 'the medicalisation of science news' in the British press in the latter half of the 20th century, and Pellechia (1997) found that a set of elite US newspapers focused on medicine and health in more than 70 per cent of their science-based stories during the same period. Of the 47 newspaper science sections identified in a survey of US newspapers by the Scientists' Institute for Public Information in 1992, most emphasised health (Jerome 1992) and Einsiedel (1992) encountered a similar dominance of health topics in an analysis of science stories in seven Canadian newspapers. Television is a more eclectic medium in most countries, with an often strong focus on natural history and environmental issues. But here, too, medicine and health often dominate (Gregory and Miller 1998).

In a study of science coverage in a leading Italian newspaper over 50 years, Bucchi and Mazzolini (2003) also found that biology and medicine accounted for more than half of the stories. But they noted that the medicalisation of science was particularly pronounced in stories written for the newspaper's special supplements and sections, while science news featured on the front page was dominated by physics and engineering stories. This suggests that science journalists may be making a conceptual distinction between news and 'news you can use', the latter focusing more heavily on health and medicine topics.

Television science reinforces the legitimacy and sacredness of science

Analyses of science television venues are less common, and generally focus on the entertainment aspects of scientific discoveries and processes and a corresponding dearth of in-depth and critical treatment (Metcalfe and Gascoigne 1995; LaFollette 2002). Collins (1987), in an analysis of a set of British documentary science programmes, noted the heavy overlay of certainty that accompanied the programmes. 'Television', he concluded, 'presents science as producing unambiguous and intractable knowledge' (ibid. 709).

Drama plays a major role in much of science television programming and, according to scholars, can often trump public understanding of science goals. Silverstone (1984, 1989) studied the making of a science documentary for the BBC *Horizon* series, and tracked the gradual takeover of the storyline by the television producers. Although the scientific information came to the process with its own strong discourse elements, television reacted with equally strong and dramatic discourse needs. Ultimately, noted Silverstone, it was television's needs that were met by the final product.

This does not mean that television strips science of its dignity. It is critical for science journalists to maintain the status of science in their dissemination processes, as that status legitimises not only their content, but also the reporters themselves. Thus Hornig's (1990) examination of several documentaries broadcast in the US programme series *Nova* found that the programmes maintained the 'sacredness' of science by portraying scientists as special and distinct from other professions and social roles.

The role of science journalists in local television news is particularly difficult to study, as these programmes are rarely archived. Researchers have taken advantage, however, of the recent creation of a massive archive of local news programmes from all over the USA. In an analysis of health stories from a representative sample of the top 50 US media markets, Pribble et al. (2006) found that, while local television stations devote significant airtime to health stories, many of the topics that were emphasised (West Nile virus, for example) were insignificant causes of mortality compared with topics that were ignored. Put another way, they found a bad fit between topics highlighted in local news stories and those health problems most likely to afflict viewers.

Coverage of science follows journalistic rather than scientific norms

Media coverage of science looks a lot like coverage of other arenas, principally because the primary drivers of coverage patterns are not the content areas on which stories are focused but, instead, the production infrastructure through which that content must pass.

For example, science stories – like all journalistic accounts – tend to be episodic in nature. Journalists are more likely to produce shorter stories about concrete happenings than longer, thematic stories about issues. Behind this is the rapid pace of most media production processes. Daily or, in the case of radio and internet news sites, hourly production cycles cannot wait for months-long scientific processes to spool out. Journalists produce stories about pieces of processes, and hope that faithful readers will be able to knit together a larger picture from these bits of fabric.

Episodic coverage does not lend itself well to discussions of process. So, not surprisingly, analyses of science stories find few descriptions of the research methods employed. Dimopoulos and Koulaidis (2002) found that nearly 75 per cent of the science stories they analysed from four Greek newspapers contained no reflection on the methodological 'how' of the scientific process, and discussion of that dimension in the remaining stories was brief and superficial. Einsiedel (1992), too, noted that most of the Canadian stories her team analysed virtually ignored process details. And a study of the coverage of scientific research in Dutch newspapers (Hijmans et al. 2003) found, similarly, that most stories eschewed complex process information.

Science journalism, again in ways typical of other types of journalism, seeks to hang stories on traditional news pegs, characteristics of real-world processes that are proven audience attention-getters. Among those pegs are characteristics such as timeliness, conflict and novelty. For example, rather than dip into a scientific research process at some haphazard stage, the science journalist waits until the completed work is on the cusp of publication in a scientific journal. That moment of publication offers a prized timely angle, an opportunity to grab the attention of a reader or viewer with the words: 'In today's issue of *Nature* . . . '.

These moments also tend to coincide with points in a process recognised – designated, if you will – as salient by the scientific culture. Journalists typically 'buy' into the legitimising structures of sources (Fishman 1980), uncritically accepting sources' designation of what is important and worthy of notice. Scientists can easily sell the argument that journalists must respect scientific process and, for example, wait for peer review to take place before embarking on a wider dissemination of research results. Scientists often complain that journalists pay undue attention to mavericks and outliers, but studies of media coverage of contested science suggest that those stories overwhelmingly reflect mainstream points of view (Goodell 1986; Nelkin 1995).

This reliance on news pegs also means that coverage of a long-running issue waxes and wanes with the presence or absence of pegs. Scientists and policy-makers will struggle for decades, for example, to understand the mechanisms of cloning and to develop sane means for society to adapt to the technique's many tantalising and alarming possibilities. But coverage of the issue will erupt only when 'something happens' in a journalistic sense, when a prime minister formally announces a new initiative, when a team of scientists unveils the first cloned cat, when a religious group lodges a complaint. While the disjunction between coverage and process can be disconcerting to some scientists, others have learned to take advantage of reporter dependence on news pegs and have become facile at guiding coverage. For example, if an important paper is about to be published in a journal, scientists may hire consultants to help them 'market' their discovery to the press by appealing to the demand for news pegs. The resulting press conferences and reporter 'exclusives' may be more influential in generating coverage than the original papers themselves.

The most important audiences in terms of shaping journalists' practice are their editors and their sources. While their 'real' audience – members of the public – have only marginal access to the newsroom, science writers are in daily touch with their sources and bosses. Thus coverage is more likely to be responsive to the priorities of these individuals. This may seem unpersuasive to scientists who feel that journalists often run roughshod over them and treat their information cavalierly. But studies of media coverage of science have demonstrated repeatedly that the scientific culture is a powerful driver of what becomes news about science. Sociologist Dorothy Nelkin, in her seminal book *Selling Science* (Nelkin 1995), reflects that media science stories overwhelmingly represent scientists as successful problem-solvers, as white knights who ride to the rescue as the world warms at an unprecedented rate, as the incidence of autism rises sharply and mysteriously, as energy sources are depleted. Such coverage is not accidental, she notes; the scientific culture actively cultivates its image as society's major driver of uncertainty reduction.

Two long-standing journalistic norms – objectivity and balance – have come under intense scrutiny in recent years. Both arose as surrogates for validity – as ways of compensating for journalists' inability to determine whether a source's assertions are true or not. They are particularly salient in science journalism, as much of science is contested terrain. What is a journalist to do when credible scientists make contradictory claims about a particular issue? The default response is to adopt a stance of objectivity and balance (Dunwoody 1999).

In a world where the science journalist cannot determine what is true, objectivity demands that the reporter go into 'neutral transmitter' mode and focus not on validity, but on accuracy. That is, rather than judging the veracity of a truth claim, the journalist concentrates instead on representing the claim accurately in her story. The issue is no longer whether the claim is supported by evidence but, rather, the fit between what a source says and what a journalist presents.

Similarly, when a science reporter cannot determine who is telling the truth, the norm of balance suggests that he represent as many truth claims as possible in the story.

When validity is impossible, in other words, a good fallback position is comprehensiveness. The journalist is, in effect, telling the reader: 'the truth is in here somewhere'.

Scholars contend that objective, balanced stories about science are misleading. Boykoff and Boykoff (2004), for example, argue that balance too often means giving truth claims equal space, even when they are not, in fact, equally valid. They use the example of global warming coverage in US newspapers to demonstrate that, even in the face of burgeoning consensus among scientists that humans are making a substantial contribution to warming, many media accounts still give significant play to global warming outliers who dispute the trend. Mooney and Nisbett (2005) find similar patterns in coverage of the debate in the USA over teaching evolution in the biology classroom; attempts to 'balance' the arguments of biologists and creationists, they claim, confer legitimacy on both sides in the minds of readers.

At least one American study indicates that journalists are keenly aware of the problems created by objective, balanced accounts, but feel that journalistic norms prevent them from abandoning these behaviours. Dearing (1995) found the expected balancing of extreme points of view in coverage of several scientific issues where strong majority viewpoints were being contested by outliers. In interviews, the journalists readily acknowledged the probably bogus nature of the mavericks' positions, but then indicated that both editors and audiences expected their stories to treat those positions with respect.

Another contested arena: the accuracy of science news stories

Both scientists and science journalists place a large premium on the accuracy of science stories. But they disagree on the nature of that evaluation, with scientists charging that much science news is inaccurate, and journalists contending that their work will stand the test of time. Resolving this disagreement is difficult, as accuracy, like bias, is a judgement that we all reach through filtered lenses.

For example, accuracy studies that ask the sources to identify inaccuracies in media science stories find that those perceived flaws are overwhelmingly errors of omission (details of study methods missing, limitations of findings omitted, names of co-authors omitted) and that scientist-sources typically reach those conclusions with readers other than the public – their peers, for example – in mind (Dunwoody 1982). Journalists respond that their general readership will be intolerant of the level of detail desired by scientists, that the typical audience member ingests a science story rapidly and superficially, making precision much less important than accurately conveying one or two take-home messages.

At least one study seems to back up this latter claim. In the early 1970s, mass communication researchers at the University of Minnesota designed a very different study of science news accuracy. Instead of sending the published story back to the scientists for an accuracy check, the researchers gave the story to individuals to read, asked those individuals to write a summary of what they had learned, and then sent the readers' summaries back to the scientists for an accuracy check. The result? Scientists judged two-thirds of the summaries to be accurate (Tichenor et al. 1970).

Scientists' tendency to view their peers as primary audiences for mediated accounts may influence their assessment of journalists' treatments, but it also has merit. Although scientists denigrate suggestions that they rely on mediated channels for information about science, evidence suggests otherwise. Like decision-makers in other realms, researchers rely on the mass media as important alerting mechanisms. Research that gets wide media attention 'looks' more important, not only to society but also to other scientists (Phillips et al. 1991; Kiernan 2003).

Appropriate training for science journalists remains a contentious topic

Should science writers be formally trained in science, or should they more properly come up through the ranks of journalism? If one looks across countries, the former increasingly seems to trump the latter. In some countries, doctoral degrees are highly sought by newsrooms; in others, science writing training programmes increasingly favour those applicants who have science credentials. The argument embedded in these preferences is not that journalistic training is irrelevant, but that a marriage of scientific and journalistic skills will yield better results than will journalistic skills alone.

The value of formal science training seems obvious and, not surprisingly, is strongly endorsed by the scientific culture, which feels that such grounding will produce more accurate and responsible stories. Many science graduates in search of careers outside science find that science writing has intuitive appeal.

It is interesting to note that there is little empirical evidence to support these training beliefs. Only a few studies have been conducted in the USA to explore differences in journalistic quality that can be pinned to differences in training, and none of those studies has found formal science training to be strongly predictive of that quality. For example, Wilson (2000) gave American environmental journalists a global warming knowledge test, then compared the answers of those journalists with formal science training and the answers of those without such training. While a formal science education made a modest difference in reporter knowledge, it was trumped by another variable: number of years on the job. Years on the job has proved the best quality predictor in a variety of studies of journalistic work in the USA (Dunwoody 2004). As is the case for most skilled occupations, experiential learning is probably the most critical predictor of job performance.

Trends in science journalism

Much change is on the horizon for journalism generally, and for science journalism specifically. In this last section I reflect briefly on several trends that are already under way.

The great shift to the internet

The availability of the internet as an information channel has profoundly affected audiences' patterns of information-seeking. In many countries, traditional media channels are either in holding patterns (television) or in decline (newspapers) as the public adjust to the enormous amount of information available electronically. For example, although television still holds pride of place in the USA and Europe as the major source of science news, and in a Pew Center survey (Horrigan 2006) 41 per cent of Americans indicated that TV was the place where they got most of their science news and information, the internet is second and closing fast (20 per cent report that they employ the internet as their chief channel for science news and information). The internet is the top channel choice for Americans when they need information on a specific science topic (Horrigan 2006).

A dominant internet environment does not mean 'anything goes'. In fact, many people seek the sites of established news media on the internet, just as they do in more traditional modes; popular sites in the USA include CNN.com, MSNBC.com and NYTimes.com. There still seems to be an enduring need for a credible, initial filter on information, and historically that filter has been the journalist and her media organisation. Still, the internet will change a science journalist's environment in a number of ways:

- the channel requires not only strong narrative skills, but also equally strong visual ones; science journalists must become increasingly multimedia in nature;
- the speed of the internet will make timeliness an even greater priority in the news business; quick turnarounds do not nurture storytellers but, instead, require journalists who relish the 'signaling' capability of the business;
- the reliability and validity of science stories will come under increasing scrutiny as readers exercise their ability to seek out multiple narratives about the same topic (see Chapter 13 in this volume).

Relationships between scientists and journalists are becoming increasingly complex

As noted above, for much of the 20th century scientists were not particularly active in public contact and, as a result, knew much less about public communication processes than did the journalists who contacted them. That gave journalists a small edge in their relationships with their sources, an edge that has begun to evaporate as scientists increasingly realise the value of public visibility and take active steps to structure their own public image. Many of today's scientists come equipped with media training and, harking back to the 19th century, are communicating directly with the public on their own through popular books or websites.

These savvy scientists are changing the public communication playing field quite radically. Now both journalists and sources come equipped to interact, with their own needs and motivations for shaping their public images. A couple of decades ago, two British researchers came up with a label for this kind of space: a shared culture. Blumler and Gurevitch (1981) argued that journalists and sources who need each other (usually for very different reasons) fashion a shared space where each group can work with the other to achieve its own ends. Each group understands the norms governing the behaviour of the other, and they sometimes construct a few rules of their own for the shared space that they both inhabit.

Science journalists will need to be careful and alert in managing this shared space. On the one hand, the new spirit of cooperation among scientists can be disarming in an environment historically fraught with tension and suspicion. But on the other hand, the increased sophistication of scientists as sources may promote the scientific culture's ability to control what becomes news. Research institutions are also increasingly active in trying to shape the agenda of the media through their press offices and PR staff (see Chapter 10 in this volume). As the dynamics of this relationship change, science journalists will have to work hard to maintain their status as independent, sometimes critical, observers (Goepfert 2007).

Relationships between science journalists and their audiences are becoming increasingly complex

Historically, audiences' relationships with journalism have been superficial at best. The news production process has functioned as a one-way street, with journalists making decisions about what is news, and audiences consuming the result, ideally without complaint. Attempts by readers or viewers to influence the process have typically been met with resistance, journalists even going so far as to decry such efforts as a form of tampering, of pre-publication censorship.

Now that journalism needs to care deeply about the needs and interests of its audiences – the decline in readership and viewership of mainstream mass media products have journalists scrambling to seek solutions – science journalism, along with the rest of the industry, is now taking some first baby steps towards a more interactive relationship with audiences. The internet is fostering this relationship, and journalists will become increasingly adept at understanding readers' and viewers' needs, and adapting to them. The trick, as always, is to become more responsive while still exercising the kind of news oversight that has made science journalism so valuable in the first place: to cater to audience interests while still providing the kind of science information that a complex polity needs in order to make sense of the world around it.

Suggested further reading

Blum, D., Knudson, M. and Henig, R. M. (eds) (2006) *A Field Guide for Science Writers*, 2nd edn, Oxford: Oxford University Press.

Burnham, J. C. (1987) *How Superstition Won and Science Lost: Popularizing Science and Health in the United States*, New Brunswick, NJ: Rutgers University Press.

Friedman, S. M., Dunwoody, S. and Rogers, C. L. (eds) (1999) *Communicating Uncertainty: Media Coverage of New and Controversial Science*, Mahwah, NJ: Erlbaum.

Gregory, J. and Miller, S. (1998) *Science in Public: Communication, Culture, and Credibility*, Cambridge, MA: Basic Books.

LaFollette, M. C. (1990) *Making Science Our Own: Public Images of Science 1910–1955*, Chicago, IL: University of Chicago Press.

Nelkin, D. (1995) *Selling Science: How the Press Covers Science and Technology*, 2nd edn, New York: WH Freeman & Co.

Other references

Bauer, M. (1998) 'The medicalization of science news – from the "rocket–scalpel" to the "gene–meteorite" complex', *Social Science Information*, 37: 731–51.

Blum, D., Knudson, M. and Henig, R. M. (eds) (2006) *A Field Guide for Science Writers*, 2nd edn, Oxford: Oxford University Press.

Blumler, J. G. and Gurevitch, M. (1981) 'Politicians and the press: an essay on role relationships', in Nimmo, D. D. and Sanders, K. R. (eds) *Handbook of Political Communication*, Beverly Hills, CA: Sage, 467–93.

Boykoff, M. T. and Boykoff, J. M. (2004) 'Balance as bias: global warming and the US prestige press', *Global Environmental Change*, 14: 125–36.

Bucchi, M. and Mazzolini, R. G. (2003) 'Big science, little news: science coverage in the Italian daily press, 1964–1997', *Public Understanding of Science*, 12: 7–24.

Burnham, J. C. (1987) *How Superstition Won and Science Lost: Popularizing Science and Health in the United States*, New Brunswick, NJ: Rutgers University Press.

Collins, H. M. (1987) 'Certainty and the public understanding of science: science on television', *Social Studies of Science*, 17: 689–713.

Dearing, J. W. (1995) 'Newspaper coverage of maverick science: creating controversy through balancing', *Public Understanding of Science*, 4: 341–61.

Dimopoulos, K. and Koulaidis, V. (2002) 'The socio-epistemic constitution of science and technology in the Greek press: an analysis of its presentation', *Public Understanding of Science*, 11: 225–41.

Dunwoody, S. (1982) 'A question of accuracy', *IEEE Transactions on Professional Communication*, 25: 196–9.

—— (1999) 'Scientists, journalists and the meaning of uncertainty', in Friedman, S. M., Dunwoody, S. and Rogers, C. L. (eds) *Communicating Uncertainty: Media Coverage of New and Controversial Science*, Mahwah, NJ: Erlbaum, 59–79.

—— (2004) 'How valuable is formal science training to science journalists?'. *Communicacao e Sociedade*, 6: 75–87.

Einsiedel, E. F. (1992) 'Framing science and technology in the Canadian press', *Public Understanding of Science*, 1: 89–101.

Fishman, M. (1980) *Manufacturing the News*, Austin, TX: University of Texas Press.

Friedman, S. M., Dunwoody, S. and Rogers, C. L. (eds) (1999) *Communicating Uncertainty: Media Coverage of New and Controversial Science*, Mahwah, NJ: Erlbaum.

Goepfert, W. (2007) 'The strength of PR and the weakness of science journalism', in Bauer, M. and Bucchi, M. (eds) *Journalism, Science and Society: Science Communication between Journalism and Public Relations*, London and New York: Routledge.

Goodell, R. (1977) *The Visible Scientists*, Boston, MA: Little, Brown.

—— (1986) 'How to kill a controversy: the case of recombinant DNA', in Friedman, S. M., Dunwoody, S. and Rogers, C. L. (eds) *Scientists and Journalists: Reporting Science as News*, New York: Free Press, 170–81.

Gregory, J. and Miller, S. (1998) *Science in Public: Communication, Culture, and Credibility*, Cambridge, MA: Basic Books.

Hijmans, E., Pleijter, A. and Wester, F. (2003) 'Covering scientific research in Dutch newspapers', *Science Communication*, 25: 153–76.

Hornig, S. (1990) 'Television's *NOVA* and the construction of scientific truth', *Critical Studies in Mass Communication*, 7: 11–23.

Horrigan, J. B. (2006) *The Internet as a Resource for News and Information about Science*, Pew Internet and American Life Project. http://www.pewinternet.org

Kiernan, V. (2003) 'Diffusion of news about research', *Science Communication*, 25: 3–13.

—— (2006) *Embargoed Science*, Urbana, IL: University of Illinois Press.

Jerome, F. (Fall 1992) 'For newspaper science sections: hard times', *SIPIscope*, 20: 1–9.

LaFollette, M. C. (1990) *Making Science Our Own: Public Images of Science 1910–1955*, Chicago, IL: University of Chicago Press.

LaFollette, M. (2002) 'A survey of science content in US broadcasting, 1940s through 1950s', *Science Communication*, 24: 34–71.

Metcalfe, J. and Gascoigne, T. (1995) 'Science journalism in Australia', *Public Understanding of Science*, 4: 411–28.

Mooney, C. and Nisbet, M. C. (2005) 'Undoing Darwin: as the evolution debate becomes political news, science gets lost', *Columbia Journalism Review*, September/October: 31–9.

Nelkin, D. (1995) *Selling Science: How the Press Covers Science and Technology*, 2nd edn, New York: WH Freeman & Co.

Pellechia, M. G. (1997) 'Trends in science coverage: a content analysis of three US newspapers', *Public Understanding of Science* 6: 49–68.

Phillips, D. P., Kanter, E. J., Bednarczyk, B. and Tastad, P. L. (1991) 'Importance of the lay press in the transmission of medical knowledge to the scientific community', *New England Journal of Medicine*, 325: 1180–3.

Pribble, J. M., Goldstein, K. M., Fowler, E. F., Greenberg, M. J., Noel, S. K. and Howell, J. D. (2006) 'Medical news for the public to use? What's on local TV news', *American Journal of Managed Care*, 12: 170–6.

Silverstone, R. (1984) 'Narrative strategies in television science: a case study', *Media, Culture and Society*, 6: 377–410.

—— (1989) 'Science and the media: the case of television', in Doorman, S. J. (ed.) *Images of Science*, Aldershot: Gower, 187–211.

Tichenor, P. J., Olien, C. N., Harrison, A. and Donohue, G. (1970) 'Mass communication systems and communication accuracy in science news reporting', *Journalism Quarterly*, 47: 673–83.

Wilson, K. (2000) 'Drought, debate, and uncertainty: measuring reporters' knowledge and ignorance about climate change', *Public Understanding of Science*, 9: 1–13.

3

Science museums and science centres

Bernard Schiele

This chapter reviews the historical development of science communication through science museums and science centres, over several centuries. It records how the development of scientific practice was embedded in, and grew from, the exhibition practices of the museum, and how changing concepts of the place of science in society are reflected in science museums and science centres. It shows how increased attention to communication through the science museum or science centre has eventually brought the visitors to centre stage.

Antiquity, the Middle Ages and the Renaissance

It is common, and historically well founded, to attribute the origin of the term *mouseion* to the name of the Greek temple built on Mount Helikon in Athens, devoted to the Muses, and harbouring a preserved treasure. In antiquity, two driving forces, piety and cupidity, spurred the gathering of works of art. The custom of *ex voto* meant that treasures were entrusted to the temple, while the plunder and spoils of war also served to enrich it (Pomian 1978). The cathedrals, churches and sanctuaries of the Middle Ages continued this tradition.[1] The modern museum that emerged at the end of the Renaissance united these two principles of accumulation and made the art object a sacred treasure, transposing it from a medium of worship to an object of worship (Deloche 1985).

The science museum selects what merits inclusion in its collection according to a different logic. During the Italian Renaissance, the *galleria* referred to a long reception hall, where works of art served a solely decorative function. But the *gabinetto* was a place reserved for the learned scholars. It was a square room, with shelf-lined walls displaying a vast array of natural and artificial curiosities for their artistic, mechanical or scientific value. This packed jumble of skeletons, fossils, mechanical automata, vegetable and mineral specimens, paintings, archeological objects, weapons, clocks, etc., would long be the museum norm (Mauriès 2002).

The Enlightenment and scientific thinking

In the 16th century, the cabinet of curiosities sought a continuity between art and nature, the principle of unity, by looking for 'similarities' and 'resemblances' in art and nature (Foucault 1970). It sought to unite the marvels of this transformation of art into nature, and nature into art, to better reveal the avowed correspondence between man and nature. From the end of the Renaissance to the beginning of the 20th century, the emancipation of reason accompanied by the rise in scientific thinking led to the modern science museum – which self-invents and evolves in keeping with the science that constitutes it.

At the turn of the 17th century, collections once reserved solely for the Prince at court and for learned scholars were opened as museums to the public. In the 18th century, those intent on understanding the natural laws governing the universe, and aligning humanity with it, took pains to gather and preserve evidence: natural specimens and man-made creations, artistic and scientific. From this sprang the idea of showing these to the public for its own edification (Laissus 1986). The origins of the museum, the functions of conservation and education, became linked: to show what is worthy of being conserved in order to educate. The first university museum was founded at Basel in 1671, the Ashmolean Museum at Oxford in 1683, then the British Museum in 1753 following the purchase by Parliament of Sir Hans Sloane's collection of natural sciences, and the Louvre in 1793, soon after the French Revolution (Poulot 1997).

The Ashmolean Museum is the first true museum in the modern sense. It participated in, and contributed to, the movement to emancipate reason. The desire to understand the universe and humanity's place in it fitted perfectly with a very profound change: liberation from the tutelage of religious thinking. At the source of the rise of reason was the dissociation – applied in the Renaissance – of the opposition between the finite and the infinite. Infinity did not negate the finite; reason could act freely in all directions to think of reality (Laïdi 1999). Scientific thinking was developing to the point of articulating a will to understand the universe with faith in a reason that is 'one and identical for any thinking subject, nation, era, or culture' (Cassirer 1997: 41). The four major principles that define the scientific method flow from this: reality exists independently of the representations we make of the world; reality is independent of the language used to describe it; the truth is a precise representation; and knowledge is objective (Searle 1995).

The museum tradition, particularly the museology of science, rests on this intellectual foundation. Traditionally, its mandate has been to guarantee, purvey and propagate it. The encyclopaedic spirit at the end of the 18th century reinforced and entrenched this mandate. And despite the inevitable transformations of museum premises, the forms adopted or the functions they fulfilled, during the 19th and 20th centuries the reference knowledge governing their expression, and the consequent exhibiting of collections, have always come out of the scientific community (or for art museums, the artistic community) and been controlled by it.

Natural science museums

From the specialisation of the sciences came the specialisation of the museum itself throughout the 19th century. With the rise of rationality, the era of accumulation progressively ended. The disorder of the cabinet of curiosities demanded ordering and classification.

The inquiry into nature quickly transposed collections into what we now call research centres. Fossils, botanical and mineral specimens, preserved animals and skeletons are the essential materials that interested scholars, who observed and compared them (Foucault 1970). These collections laid the foundation for emerging scientific thinking, which, in turn, brought systematisation. Zoology, botany, geology, paleontology, comparative anatomy – all these disciplines, destined to dominate the scientific field in the 19th century, were already a latent presence in the disparate assortments of the cabinets of curiosity. This trend was bolstered by the richness of most of these collections, amassed by wealthy and impassioned amateur collectors, and the care taken in classifying and organising them (Pomian 1987).

The work of classification changed as the fields of knowledge were defined. This evolution led to a rejection of groupings based on the previous analogies and correspondences, and instead gave prominence to those based on methods developed by the fledgling natural sciences. The 'naturalia' were grouped in natural history museums, and dissociated from the 'artificialia' placed in art museums.[2]

The preoccupation throughout the 18th century was the classification and ordering of nature. In 1773 Carl von Linné (Linnaeus) (1707–78) proposed in *Species plantarum* a method of binomial classification that was quickly adopted by the naturalists. Exhibiting of the natural sciences was then identical to systematic collection, and merged seamlessly with it (Van-Praët 1991). This would continue until the end of the 19th century in the museums where researchers were carrying out essential scientific work. The opening of the Pitt Rivers Museum in Oxford in 1883 introduced a distinction between exhibiting itself, and the content of the collections. Abandoning the classification approach for a synthetic perspective, it prefigured the future evolution of museums, where objects are presented according to a typological relationship rather than by the geographical origin of ethnological and anthropological collections (Pearce 1989).

Three mutually reinforcing factors explain this change in exhibiting style. The first was the considerable growth of collections. The 19th century is, par excellence, the century of the natural sciences – the planet was widely explored through expeditions monopolised by Western explorers. The second factor was the development of the sciences themselves. A synthetic perspective followed the classifying period, symbolised by Darwin's *On the Origin of Species* (1859). Finally, the desire to educate the public meant including communication of the sciences as part of exhibiting. The tension created by this push for communication was apparent during the founding of the Smithsonian Institution, notably in the resistance of Joseph Henry (its first secretary) to any form of activity other than research. He felt strongly that producing knowledge should take precedence over its communication (Oehser 1970). However, Henry could not forestall the development of what would become the National Museum of Natural History.[3] The shift towards the logic of communication (even though the word was not yet in vogue) was accelerated by the development of dioramas throughout the 19th century (Altick 1978). With the diorama, the objective was to attract, captivate and retain the visitor's attention by making nature spectacular (Wonder 1993).

From cabinet of physics to science and technology museum

The science and technology museum has three lines of descent. The first relates directly to the emergence of scientific thought, which brought the natural sciences to centre stage throughout the 19th century. The second is the development of cabinets of physics and chemistry. These are distinguished from the cabinets of natural science by their focus on experiments. Beginning in 1730 in France, Abbé Nollet (1700–70), an unrivalled populariser who was elected to London's Royal Society, made 'experimental physics an amateur pleasure and a fashionable entertainment' (Torlais 1987). He created a public for physics by demonstrating spectacular experiments, and by creating a dialogue with the public that combined curiosity and reason in his demonstrations. The cabinets of chemistry, while definitely less spectacular, served to purvey the new chemistry of Priestley (1733–1804) and Lavoisier (1743–1794) as fast as they could invent them. These cabinets also housed machines and mechanical automata, evidence of a close association between science and technology. From their very beginnings, museums of science and technology chose 'what was new, what could most readily be demonstrated by experiments, and would then explore the state of the question and add everything possible from the arts and from machines' (Torlais 1986: 623). This taste for the spectacular at the service of technology and industry was magnified by the gigantism of world exhibitions, the third line of descent of science and technology museums.

The first technology museum initiative was the outcome of the Descartes project to establish a Conservatoire National des Arts et Métiers (created in 1798) to explain the making and use of tools and machines, to promote the technical and commercial arts, and to encourage industrial development. But, in hindsight, it appears that the real driving force behind science and technology museums was the world exhibitions that accompanied the industrial revolution. In this euphoric period of technical progress, the world exhibitions celebrated the achievements of the human mind and sought to show how art and science could contribute to producing industrial objects. They illustrated how the many applications of the machine had enhanced the lot of human beings and contributed to their wellbeing. Driven by this conviction, the first international exhibition, the Great Exhibition of the Works of Industry of all Nations, held in London in 1851, compared the industrial progress of various nations (Greenhalgh 1988; Schroeder-Gudehus and Rasmussen 1992; Rydell 1993, 2006).

Once these exhibitions were over, the objects, and sometimes even the buildings erected for the event, served to establish new museums. The Great Exhibition, familiarly called the Crystal Palace, engendered the Science Museum of London, which houses that collection. The Technical Museum of Vienna reused the material from the Vienna International Exhibition of 1873. The Smithsonian inherited presentations from the Philadelphia Centennial Exposition of 1876. The Exposition Internationale de l'Electricité of 1881 in Paris inspired Oskar von Miller, founder of the Deutsches Museum in Munich. The Palais de la Découverte in Paris, built on the occasion of the International Exhibition of 1937, was transformed into a permanent establishment in 1938.

But the world exhibitions went beyond simply creating or enriching the collections of certain museums. They were founded on the relationship that these museums wanted to have with their visitors. Basically, they wanted to teach. They also wanted to be democratic and open to everyone, to be comprehensible in every aspect, with guided tours and lecture tours organised in a number of languages. Teaching was intrinsic to the exhibition: the Swedish Pavilion at the Vienna Exhibition (1873) presented a model school. Besides this pedagogic mission, these exhibitions also wanted to be more entertaining, with recreational or play-oriented aspects to grab visitors' attention. These three features: links to the schools, dramatic and spectacular presentations (theatricality), and the desire to educate while entertaining, would have a decisive impact on the development of science and technology museums, and represent their principal features today.[4]

All this was based on a constant concern inspired by the large-scale stores that were opening at that time: how best to use the space for installation of the objects and to define circulation patterns that would enable visitors to see everything, to be attracted and captivated more-or-less effortlessly on their part. In addition to this concern with the manner of exhibiting objects, there was concern about how the exhibited objects were to be classified. Faced with the impossibility of creating satisfactory groups – as various objects fell into several categories – and experiencing the same problem, the world exhibitions finally opted for a synthetic approach. They abandoned classification and adopted a thematic ordering of items. The Chicago Exhibition of 1933 was the first to do this systematically. These two aspects, optimisation of circulation patterns and thematic grouping, characterise the displays at contemporary science and technology museums.

The world exhibitions glorified the technical and industrial applications of science. The machinism of the 19th century also lent itself to a spectacular show where pure science, considered cold and arid, was generally absent. Only with the Paris Exhibition of 1878 was pure science to be truly celebrated for itself. By the turn of the century, scientific progress was revolutionary, and the invention of electrical lighting, the telephone and the phonograph spurred expectations of a radical transformation in lifestyle. Science was in the forefront, with the Exhibitions of Paris (1900), the Faraday exhibition (London 1931) and the Chicago exhibition entitled A Century of Progress (1933). This focus on science was reflected in the creation of the Palais de la Découverte on the occasion of the 1937 Exposition in Paris, which was devoted entirely to basic research.

At the time of its opening, the Palais de la Découverte in Paris was the jewel of French science museology. It was the ultimate in modernity.[5] What were its salient features? Jean Perrin (1870–1942) wrote at the time:

> What we wanted was to familiarise our visitors with the basic research whereby science is created, by repeating on a daily basis the great experiments that generated this research, without reducing the level but keeping them accessible to a wide range of thought. In so doing, we wanted to engender a public taste for scientific culture along with the qualities of precision, critical integrity and freedom of judgment which this culture develops and which are useful for everyone, no matter what their career.
>
> (Quoted by Rose 1967: 206)

Like all museums and science centres that have opened since, the Palais de la Découverte assumed a mission of communicating and sharing scientific culture. It sought to introduce scientific thinking, and in order to have people 'understand the determining role of Discovery in the creation of civilisation', it produced living exhibitions 'demonstrating spectacularly as much as possible the basic discoveries that have expanded our intelligence ... assured our command over matter ... or increased our physiological security' (Perrin, quoted by Roussel 1979: 2). Thus the affirmation of a social necessity for science took root in the profound conviction that pure research,[6] a quasi-aesthetic activity motivated by free curiosity, is a disinterested quest of the unknown that leads to discovery. This perspective was shaped by the Palais de la Découverte through the types of mediation it set up: the exhibition–show, the demonstrators, the invitation to touch and the push button – the latter two heralding interactivity (Eidelman 1988).

The science centre is organised entirely around the disciplinary knowledge and fundamental sciences that it promotes and purveys. Its model of mediation rests on the classroom–laboratory transposed into an exhibition–show, animated by demonstrators who reproduce spectacular experiments and explain them to an audience. It 'gives to the individual's perceptions of the outside world a viewing screen of concepts upon which to project and refer one's perceptions' (Moles 1967: 28). The Preamble to the Palais de la Découverte project states: 'demonstrators (with phonographic discs and cinematographic films) will give the necessary explanations. Brief comments in the form of pictures logically relate the experiments and form a logical whole for each science, indicating the inventions or practical applications that accrue from each discovery' (Eidelman 1988: 10).

This summary of key ideas from the Palais de la Découverte clearly shows that society puts its stamp on the museum: it is an outcome of a certain concept of science, and the production devices and uses of the dominant knowledge. This vision, echoed by Jean Perrin, that basic disinterested research, 'through a singular return, is at the source of all inventions of a useful nature' (Eidelman 1988: 45), dominated most of the 20th century. The programme is set forth in the celebrated report by Vannevar Bush, *Science, The Endless Frontier*, written at the end of World War II, which led to the apparatus of American basic research supported by the state and concentrated in the large universities (Bush 1960).

From science and technology museum to science centre

From the 1970s, the centre of gravity shifted, corresponding to the rise of the technosciences (Papon 1989), and bringing changes to the museum institution. The advent of science centres, purposefully opening a window on technical and industrial knowledge, characterised this reconfiguration. The opening in 1969 of the Exploratorium in San Francisco (Hein 1990) and the Ontario Science Centre in Toronto exemplified this reversal of trends. Five factors are to be considered: communication shift, interactivity, evaluation, risk and environmental concerns, and spin-offs of science.

These two science centres were the first resolutely to make communication with visitors their primary objective. Until then, the science had been the prime focus of attention, and communication was a tool to serve scientific knowledge; from then on, communication would take precedence, with the intention of raising interest in science and helping to achieve science literacy. This shift denoted a complete reversal of trend.

Why communication? Because changes were happening by the start of the 1970s, reflected in the radical transformation of museums which realigned their mission and transformed their practices. The first change was an accelerated movement to legitimate public communication of science and technology as a distinct practice. Popularisers, previously auxiliary to the scientific community, demanded to be the exclusive mediators with the general public. They justified this demand by denouncing the inability of scientists to address the public, to discern its expectations or be able to share the 'immense powers that knowledge brings'. While they had contributed markedly to the dissemination of scientific thinking and spirit in the 19th and early 20th centuries, they now needed a 'mediator', neither scientific nor lay, who would serve as intermediary between the scientific community and the general public (Moles and Oulif 1967). Thus each criticised the other, and in so doing, filled the knowledge gap between science and ordinary knowledge. This carried over to museums and science centres in the primacy accorded to communication, as manifested mainly in interactive exhibits.

The second change, a corollary to the first, relates to this self-same interactivity, and began to be felt in the 1970s. A report, *To Improve Learning*, published in the USA (Anon. 1970), set out how educational technology (mainly audiovisual as the electronic highway did not yet exist) could, if used appropriately and integrated systematically into teaching, and supported by research into learning and communication, increase the effectiveness and productivity of schools.[7] The idea of the 'facilitator' replacing the teacher emerged for the first time. Educational technology, although not successful in transforming the American school, nonetheless opened the way to a stronger rapprochement between the school and the museum: it would become the laboratory and the window of the transformed school. Two premises must be recalled: the need to establish a link between the school and the outside world from which it had been cut off for too long, to motivate students to respond better to their individual interests; and the need to address individual needs by adapting to differences in their diverse ways of learning. The so-called active, alternative, community, cooperative, etc. learning movement was part of this evolution. Their common thrust was to propose a pedagogic relationship where the student is central and masters his or her own learning process.

The third change, the development of evaluation in museums (Schiele 1993), with C. G. Screven as promoter and defender (Samson and Schiele 1989), is the manifestation of the museum's spotlight on the visitor, using interactivity to optimise communication. Screven maintained that the exhibition must be conceived for the visitor according to predetermined objectives. He proposed a series of measures to optimise the communication relationship set up by the exhibition and to reinforce the museum's educational role. During the 1970s, science centres were revamped in a pedagogic mould, geared to the visitor and based on a closer relationship with schools, representing the most innovative forms without being assimilated by them (Miles 1982; Loomis 1987; Bicknell and Farmelo 1993).

By making the museum a place of knowledge mediation, by soliciting visitors' active participation, voluntarily situated at the centre of the exhibit, a vision was imposed that revolutionised the practice of science museology. The success of the formula, which is still valid today, contributed greatly to the entrenchment of the concept. There is nothing exceptional about this visitor-centred polyvalence. It is characteristic of what contemporary museums offer, be they art, civilisation, society or science museums. All museums try to diversify their exhibitions, programmes, activities and the events they organise in order to create a multi-sensory and multi-communication environment that constantly attracts the visitor and increases potential anchoring points. Everything is geared to attract, stimulate, captivate and hold their attention, to motivate, awaken their interest or mobilise their cognitive styles. More than ever, museums have adopted the precepts of McLuhan, who maintained that all means used must contribute to realising the objectives of communication (McLuhan et al. 1969). Today he would no doubt add multimedia and the internet to this gamut of means.

The fourth change is represented in the growing public awareness that progress brings its own irritations and risks. The 1970s brought warning of the progressive invasion of science and technology, and its profound impact on daily life, work and the environment. Accidents of serious consequence, such as the *Torrey Canyon* oil spill in 1967, the Three Mile Island nuclear plant accident in March 1979, and the explosion at the pesticide plant in Bhopal, India in 1984, followed by Chernobyl in 1986 (Lecerf and Parker 1987; Friedman et al. 1999) brought home the tangible reality of risk for everyone. Added to this, and denounced by Rachel Carson (1962), were the persistent pollutants such as polychlorinated biphenyls, dioxins, furans, DDT and numerous pesticides present everywhere in the environment, the effects of which constitute serious and insidious threats to the health of populations (and the balance of ecosystems). Too many incidents raised serious questions about scientific development and the accompanying notion of progress. A systematic doubt took hold progressively, and the public became cautious and critical. People questioned the utopia of a society transformed by the reason of the Enlightenment, absorbed in the science and materialised by scientific progress. This change of sensibility was reflected in increased environmental concerns and their inclusion on the agendas of science centres (Davallon, Grandmont and Schiele 1992).

While the spirit of the Enlightenment still motivated society's relationship with science, technology and their spin-offs at the turn of the 1970s, from that time onwards there was doubt about the notion of progress associated with the Enlightenment. Progress was also the utopia of a society transformed by reason finally liberated from faiths and ideologies, and scientific progress was the adjunct to this liberation. Science museums could not escape this constraint. Visitors wanted a perspective on the issues and risks associated with the promise of science. Addressing the public means including their questioning in the museum's presentation strategies. This is achieved on the one hand by associating the ideas and the questioning, and on the other by including social issues in the proposals submitted for debate: in short, the museum must be a true public place that the public seeks.

Political discourse, along with that of interest and pressure groups, and supported, enhanced and hyped by the media, certainly contributed to a fall in science's credibility and suffered their own public setbacks. Museums, however, have escaped this public scepticism thus far: they have been perceived as credible, as public refuges, places of intersecting discourse where social actors participate and speak out. This is what visitors expect when they enter a museum with their doubts and queries. They enter with the citizen's right to speak.

But these expectations on the part of the public create tension with a fifth change, which also exerts an influence on the mission of science centres. As noted, no longer science, but the spin-offs of science, were at the front and centre. This was the case at the Cité des Sciences et de l'Industrie in Paris and, since the 1980s, at nearly all science centres that have moved progressively into this new vision. The recently opened Montreal Science Centre goes further, focusing solely on innovation. Science is referred to only occasionally, as a remote reference. If, at the Palais de la Découverte, science, the vector of progress and change, allowed others to take on the task of converting discoveries into applications, now the process was reversed: innovation replaced basic research as the engine of change (Castells 1996). The change was more profound than it first appeared. Basic research placed value on knowledge for itself: it sought to reveal the intelligibility of the world through the unlimited exercise of reason. Today it is a transformed relationship, one that has less to do with understanding the world than transforming it. Maintaining economic development, borne by relentlessly renewed consumption, demands a constant influx of new ideas and new products on the market, and knowledge kow-tows to this. Basic research is subjected to a rule of necessity that adheres to the contexts binding its validity to its potential uses.[8]

In fewer than 30 years, the relationship between society and science has changed completely, and science centres not only reflect but promote and promulgate this repositioning. Thus the image of the researcher engaged in an individual quest, as highlighted by the Palais de la Découverte, was replaced by that of teams united around a project, to which each contributes knowledge and skills. So it is a constant matter of seeking new information, upgrading knowledge, recycling (Gibbons et al. 1994; Nowotny et al. 2001). A good illustration of the new relationship with contemporary knowledge is the Science Learning Network (SLN Alliance) of the Unisys Corporation in the USA, which brings together science centres and schools so that they function as databases and resource centres for each other (Helfrich 2000). This is a relationship familiar to visitors, and something they experience every day.

The public as museum actor

Today's natural science museums advocating sustainable development or science centres showing innovations and spin-offs share the task of scrutinising the discourse of the scientific community and promoting its thinking. But the rupture and realignment of practices resulting from electronic media developments, along with mass migrations – notably the telescoping of local and global perspectives on social issues and debates (Appadurai 1996) – now highlight the extent to which the modern museum institution is rooted in society and reflects its transformations.

The museum of the 1970s made a concerted effort to be a place of communication. Of all the changes that reoriented its priorities and redefined its practices, the most determining and significant is the factor of taking the public into account. It is impossible to speculate about the museum public (who comes, who they come with, how much time they spend) without considering the situations in which they find themselves engaged (what they do there, how they behave), or the museum's efforts in presenting the configurations (physical and symbolic devices) to create the communication relationship. Museums invite their visitors to participate in activities designed for them (pedagogical activities, educational actions, animation, cultural actions, cultural dissemination), to involve them in communication relationships (learning, entertainment, reflection, observation, emotion). This systematic taking into account of the public characterises an evolution that puts visitors centre-stage (Hooper-Greenhill 1994).

Today's visitors want to be heard. What's more, they want to be actors and to exercise the right to speak. The result is a profound re-examination of the museum's mission, its traditional relationship to culture, and its generally designated mediators. Inevitably there is controversy when visitors become concerned about what the museum is doing and the ways it chooses, as shown in the debates sparked by the Science in American Life exhibition presented at the National Museum of American History, or by the project The Crossroads: The End of World War II, the Atomic Bomb and the Origins of the Cold War. Considered revisionist because it proposed a perspective on the decision to bomb Hiroshima and Nagasaki, this project was never produced – all that remains is the fuselage of the *Enola Gay* shown at the Air and Space Museum (Gieryn 1998; Macdonald 1998). A Question of Truth, presented at the Ontario Science Centre, proposed the deconstruction of scientific discourse by showing how a rationality steeped in ideology can lead to excesses – for example, all the pseudoscientific apparatus developed by the Nazis to justify their racist theories. It, too, faced strong controversy when it opened. Some saw it as a denunciation finally revealing scientism for what it was, while others saw in it an affirmation of a relativism steeped in anti-science. It seems evident that the museum's destabilised reference discourse has been called on to justify itself. Without being restricted to specialists, nor excluded from the museum's public space, it must be fully endorsed and sanctioned by those it involves before being presented to the public.

To conclude, the visitors have been redefining the rules to affirm their right to speak. But the focus on the visitor, which has progressively reconstituted the museum field over the past 30 years, was – we now know – only the surface manifestation of a much more radical transformation, reflecting the depletion of traditional sources of meaning, and bringing into question the modern institutions developed for self-perpetuation (for an in-depth discussion see Semprini 2003). The debate seldom deals with visitors any more – they have gained their central status. The debate focuses instead on the legitimacy of the intermediaries that address them, as the real issue today is the redistribution of the monopoly of legitimate speech in the cultural field.

Notes

1 The same kind of accumulation of precious objects prevailed during the same period for Islam, China and Japan (to which we probably owe the oldest museum in the world: the 8th century Shōsō-in at the Monastery of Todaiji at Nara, near Kyoto).

2 But sometimes time is needed – it was only between 1880 and 1883 that the natural science collections of the British Museum were moved and reorganised at Kensington, in what would become the British Museum (Natural History).
3 It should be emphasised that the function of dissemination of natural science museums became more important as, over the years, contemporary research in biology, microbiology, physiology, etc. shifted to the laboratory or the field, only occasionally being carried out in museums. This dissociation is clearly visible in Paris, with the Grande Galerie de l'Evolution reserved uniquely for exhibitions, while the establishment that houses it – the Muséum National d'Histoire Naturelle – is first and foremost a series of research laboratories. The Institut Royal des Sciences Naturelles in Belgium is similarly divided into different locations for dissemination and research.
4 For an overview of the history and functions of museums, see: Alexander (1987); Hudson (1987); for a critical approach see Bennett (1995).
5 The Palais de la Découverte would be radically different from the museology of the Musée National des Techniques and the Muséum National d'Histoire Naturelle, whose modes of mediation are the series.
6 We must recall how the idea of 'pure science' formed and played a part in shaping all the work of dissemination of the Palais de la Découverte. Unfortunately, this topic far exceeds the boundaries of this chapter.
7 *To Improve Learning* defines educational technology this way: 'a systematic way of designing, carrying out, and evaluating the total process of learning and teaching in terms of specific objectives, based on research in human learning and communication, and employing a combination of human and nonhuman resources to bring about more effective instruction' (Anon. 1970:19).
8 An indicator of this trend in the museum field is that museums increasingly develop their research in terms of exhibition projects, while not so long ago the reverse was the case (Lewenstein and Allison-Bunnell 2000).

Suggested further reading

Bennett, T. (1995) *The Birth of the Museum – History, Theory, Politics*, London, New York: Routledge.
Bicknell, S. and Farmelo, G. (1993) *Museum Visitor Studies in the 90s*, London: Science Museum.
Macdonald, S. (1998) (ed.) *The Politics of Display – Museums, Science, Culture*, London, New York: Routledge.

Other references

Anon. (1970) *To Improve Learning, A Report to the President of the United States by the Commission on Instructional Technology, Committee on Education and Labor, House of Representatives*, Washington: US Government Printing Office.
Alexander, E. P. ([1980] 1987) *Museums in Motion – An Introduction to the History and Functions of Museums*, Nashville, TN: American Association for State and Local History.
Altick, R. D. (1978) *The Shows of London*, Cambridge, MA and London: Harvard University Press.
Appadurai, A. (1996) *Modernity at Large – Cultural Dimensions of Globalization*, Minneapolis, MN and London: University of Minnesota Press.
Bennett, T. (1995) *The Birth of the Museum – History, Theory, Politics*, London, New York: Routledge.
Bicknell, S. and Farmelo, G. (1993) *Museum Visitor Studies in the 90s*, London: Science Museum.
Bush, V. ([1945] 1960) *Science, The Endless Frontier*, Washington, DC: National Science Foundation.
Carson, R. ([1962] 1987) *Silent Spring*, Boston, MA: Houghton Mifflin.
Cassirer, E ([1932] 1997) *La Philosophie des Lumières*, Paris: Fayard.

Castells, M. (1996) *The Network Society*, Cambridge: Blackwell.

Davallon, J., Grandmont, G. and Schiele, B. (1992) *The Rise of Environmentalism in Museums*, Québec: Musée de la civilisation.

Deloche, B. (1985) *Museologica: Contradictions et Logique du Musée*, Paris: Vrin.

Eidelman, J. (1988) 'La création du Palais de la Découverte (Professionnalisation de la recherche et culture scientifique dans l'entre-deux guerres)', PhD thesis, Paris: Université de Paris V – René Descartes.

Foucault, M. (1970) *The Order of Things: An Archaeology of the Human Sciences*, London: Tavistock.

Friedman, S. M., Dunwoody, S. and Rogers, C. L. (1999) (eds) *Communicating Uncertainty – Media Coverage of New and Controversial Science*, Mahwah, NJ and London: Lawrence Erlbaum Associates.

Gibbons, M., Limoges, C., Nowotny, H., Schwartzman, S., Scott, P. and Trow, M. (1994) *The New Production of Knowledge*, London: Sage.

Gieryn, T. H. (1998) 'Balancing acts: science, *Enola Gay* and history wars at the Smithsonian', in Macdonald, S. (ed.) *The Politics of Display – Museums, Science, Culture*, London, New York: Routledge, 197–228.

Greenhalgh, P. (1988) *Ephemeral Vistas – The Expositions Universelles, Great Exhibitions and World's Fairs, 1851–1939*, Manchester: Manchester University Press.

Hein, H. (1990) *The Exploratorium – The Museum as Laboratory*, Washington, DC: Smithsonian Institution Press.

Helfrich, P. M. (2000) 'Building on-ramps to the information superhighway: designing, implementing, and using local museum infrastructure', in Schiele, B. and Koster, E. (eds) *Science Centers for this Century*, Sainte-Foy, France, MultiMondes, 87–123.

Hooper-Greenhill, E. (1994) *Museums and their Visitors*, London, New York: Routledge.

Hudson, K. (1987) *Museums of Influence*, Cambridge: Cambridge University Press.

Laïdi, Z. (1999) *La tyrannie de l'urgence*, Québec: Musée de la civilisation.

Laissus, Y. (1986) 'Les cabinets d'histoire naturelle', in Taton, R. (ed.) *Enseignement et diffusion des sciences au XVIIIe siècle*, Paris: Hermann, 659–712.

Lecerf, Y. and Parker, E. (1987) *L'Affaire Tchernobyl*, Paris: Presses Universitaires de France.

Lewenstein, B. V. and Allison-Bunnell, S. (2000) 'Creating knowledge in science museums: serving both public and scientific communities', in Schiele, B. and Koster, E. H. (eds) *Science Centers for this Century*, Sainte-Foy, France: MultiMondes, 187–208.

Loomis, R. J. (1987) *Museum Visitor Evaluation: New Tool for Management*, Nashville, TN: American Association for State and Local History.

Macdonald, S. (1998) (ed.) *The Politics of Display – Museums, Science, Culture*, London, New York: Routledge.

Mauriès P. (2002) *Cabinets of Curiosities*, London: Thames & Hudson Ltd.

McLuhan, M., Parker, H. and Barzun, J. (1969) *Exploration of the Ways, Means and Values of Museum Communication with the Visiting Public*, New York: Museum of the City of New York.

Miles, R. S. ([1982] 1988) *The Design of Educational Exhibits*, London: British Museum (Natural History).

Moles, A. A. (1967) *Sociodynamique de la Culture*, La Haye: Mouton.

Moles, A. A. and Oulif, J.-M. (1967) 'Le troisième homme, vulgarisation scientifique et radio', *Diogène*, no. 58, 29–40.

Nowotny, H., Scott, M. and Gibbons, M. (2001) *Re-Thinking Science – Knowledge and the Public in an Age of Uncertainty*, Cambridge: Polity.

Oehser, P. H. (1970) *The Smithsonian Institution*, New York: Praeger.

Papon, P. (1989) *Les Logiques du Futur*, Paris: Aubier.

Pearce, S. (1989) 'Museum studies in material culture: introduction', in Pearce, S. (ed.) *Museum, Studies in Material Culture*, Leicester: Leicester University Press, 1–10.

Pomian, K. (1978) 'Entre l'invisible et le visible: la collection', *Libre*, no 3: 4–55.

—— (1987) *Collectionneurs, Amateurs et Curieux – Paris, Venise: XVIe–XVIIIe siècle*, Paris: Gallimard.
Poulot, D. (1997) *Musée, Nation, Patrimoine – 1789–1815*, Paris: Gallimard.
Rose, A. J. (1967) 'Le Palais de la Découverte', *Museum*, vol. 20, no. 3, 206–208.
Roussel, M. (1979) *Le Public Adulte au Palais de la Découverte (d'après les principaux résultats d'une enquête sociopédagogique, 1970–1978)*, Paris: Palais de la Découverte.
Rydell, R. W. (1993) *World of Fairs*, Chicago, IL and London: University of Chicago Press.
—— (2006) 'World fairs and museums', in Macdonald, S. (ed.) *A Companion to Museum Studies*, Oxford: Blackwell, 135–151.
Samson, D. and Schiele, B. (1989) *L'Évaluation muséale – Publics et Expositions (Bibliographie raisonnée)*, Paris: Expo-Media.
Schiele, B. (1993) 'Creative interaction of visitor and exhibition', *Visitor Studies: Theory, Research, and Practice*, Jacksonville, FL: Visitor Studies Association, vol. 5, 28–56.
Schroeder-Gudehus, B. and Rasmussen, A. (1992) *Les Fastes du Progrès – Le guide des Expositions universelles, 1851–1992*, Paris: Flammarion.
Searle, J. R. (1995) 'Postmodernism and the Western rationalist tradition', in Arthur, J. and Shapiro, A. (eds) *Campus Wars – Multiculturalism and the Politics of Difference*, Boulder, CO: Westview Press, 28–48.
Semprini, A. (2003) *La Société de Flux*, Paris: l'Harmattan.
Torlais, J. (1986) 'La physique expérimentale', in Taton, R. (ed.) *Enseignement et Diffusion des Sciences au XVIIIe siècle*, Paris: Hermann, 619–643.
Torlais, J. (1987) *Un Physicien au Siècle des Lumières: l'Abbé Nollet: 1700–1770*, Paris: Jonas.
Van-Praët, M. (1991) 'Evolution des musées d'histoire naturelle: de l'accumulation des objets à la responsabilisation des publics', in La Grande Galerie du Muséum (ed.) *La Galerie de l'Evolution, Concepts et Evaluation – Colloque international, 22–23 novembre 1990*, Paris: Cellule de Préfiguration – Muséum national d'Histoire naturelle, 19–26.
Wonders, K. (1993) *Habitat Dioramas – Illusions of Wilderness in Museums of Natural History*, Uppsala: Acta Universitatis Upsaliensis.

4

Cinematic science

David A. Kirby

Introduction

As other scholars have noted, before 2000 there were relatively few studies that examined fictional cinema as a vehicle of science communication (Weingart and Pansegrau 2003). There are numerous reasons for the dearth of work on science and cinema, including a general academic bias against popular culture. The major reason, however, for the minimal scholarship on science and cinema is the historically narrow view of the public understanding of science, the 'deficit model', which attributes negative public attitudes towards science to a lack of scientific knowledge. Under the deficit model, movies are at best an unreliable means of increasing knowledge, and at worst a medium that significantly harms science literacy by disseminating misinformation.

Science studies scholars have critiqued the 'deficit model' (Irwin and Wynne 1996), and several studies of the public understanding of science demonstrate that the meanings of science – not knowledge – may be the most significant element contributing to public attitudes towards science. According to Alan Irwin (1995), the public makes sense of science – constructs a 'science citizenship', in his terms – in the context of their everyday lives, pre-existing knowledge, experience and belief structures. Popular films influence people's belief structures significantly by shaping, cultivating or reinforcing the 'cultural meanings' of science.

Scholars have begun to recognise cinema's role in the public communication of science and technology, and its importance in the public understanding of science.[1] Although there is a need for more work to be done, there now exists an established and growing literature on science in film. The main challenge is that this literature does not emerge from a single field. These works draw on a wide variety of approaches and methodologies from numerous disciplines, including communication, sociology, history, film studies, cultural studies, literature, and science fiction studies. As with studies of science and news media, exploration of science communication in popular films revolves around four basic research questions: how is science representation constructed in the production of cinematic texts? (production); how much science, and what kind of science, appears in popular films? (content analysis); what are the cultural interpretations of science and technology in popular films? (cultural meanings); what effect, if any, does the fictional portrayal of science have on science literacy and public attitudes towards science? (media effects).

'Science' in fiction is not defined solely as factual information; it encompasses what I term the 'systems of science'. The systems of science include the methods of science, the social interactions among scientists, laboratory equipment, science education, industrial and state links, along with aspects of science that exist, in part, outside the scientific community, such as science policy, science communication, and cultural meanings. In the end, scholarship on science and cinema should be aimed at understanding how the systems of science are depicted in cinema, how these depictions have developed over time, how contemporary film-making practices contribute to these depictions, and how these depictions affect the real world systems of science. In this chapter I summarise what scholars have uncovered about science and cinema, and point towards areas that still require academic attention.

Production

Although there has been considerable work on science in the production of news media, production is the least-studied area with regard to science in cinema. Researchers into news media have analysed the interactions of key players in the production process, including scientists, journalists, editors and news organisations (Peters 1995). Scholarship in the production of cinema has focused mainly on the roles of scientists and scientific organisations in the making of films. These works argue that presentations of science in the entertainment media reveal a tension not only between the narrative forms of media and those of science, but also between the needs of the entertainment industry and those of the scientific community.

Scientists and scientific organisations working on popular films need film-makers to maintain the authenticity of scientific depictions. Film-makers, on the other hand, need only to claim authenticity for their films, and ask scientists to help them maintain an acceptable level of verisimilitude. This discrepancy in needs clearly leads to multiple interpretations of the term 'authenticity'. For scientists, authenticity requires an adherence to scientific verisimilitude throughout an entire film. Film-makers consider a film to be scientifically authentic if it has any scientific verisimilitude within the constraints of budget, time and narrative. The reason for this broad definition is that film-makers find that claims of authenticity enhance their ability to draw in audiences. Many Hollywood film-makers in the 1920s and 1930s adopted the techniques of scientifically based nature documentaries in order to claim 'authenticity' for exploitation films such as *Ingagi* (1930).[2] The artifice of exploitation films is clear. As Gregg Mitman (1999) and Derek Bousé (2000) show, however, films made by scientists, such as *The Silent Enemy* (1930), were as thoroughly constructed and inauthentic as the exploitation films.

Those making biopics and historical dramas also need to claim scientific and historical authenticity in order to be successful at the box office. But film-making practices have often hindered the pursuit of accuracy, despite film-makers' desires to maintain, or even claim, authenticity. The strictures of the Motion Picture Code, for example, placed severe limitations on how film-makers could depict Paul Ehrlich's scientific story in *Dr Ehrlich's Magic Bullet* (1940) (Lederer and Parascandola 1998). The film-makers' adherence to the Motion Picture Code meant they were not able to mention syphilis or even venereal disease, nor could they show animal experiments. 'Authenticity' was achieved in the film, but it extended only to the costumes, sets and props. Film-makers could ignore other facets of scientific and historical accuracy and still claim the film was an authentic recreation of Ehrlich's scientific life. Despite its minimal accuracy, the film proved useful to the US Public Health Service, which convinced Warner Brothers to make a 'revised' version of the film three years after its release, for educational purposes. Manhattan Project scientists participated in the making of the film *The Beginning or the End* (1947) and found that film-making practices for fictional films imposed a structure and a dynamic on scientific matters that were not anticipated by the scientists (Yandetti 1978; Reingold 1985). On the other hand, film-makers required scientists' assistance in order to maintain the concept of authenticity. To assure scientists' help, film-makers had to allow scientists a large degree of control over the final version of this film, and scientists had the power to veto any portion of the script they disagreed with.

The use of scientists as consultants on more recent Hollywood films has also been examined (Frank 2003; Kirby 2003a; 2003b). Employing science consultants represents a commodification of knowledge that allows film-makers, who have purchased this commodity, to create their own 'authenticity' with regard to science. Authenticity is not a fixed term because scientific expertise becomes a flexible concept in the context of Hollywood films. Who and what constitutes an expert is dependent on film-makers needs for their production. Ultimately, film-makers' claims of authenticity for fictional films can be harmful to science literacy because cinema naturalises both 'accurate' and 'inaccurate' science, leaving audiences either believing inaccurate information or not accepting accurate science.

Content analysis

Science communication researchers have relied primarily on content analysis of newspapers to determine what science, and how much of it, appears in news media. But very few studies of fiction have utilised this methodology outside television (Gerbner 1987). In terms of cinema, there have been only two wide-ranging, quantitative studies of science in cinema. Film scholar Andrew Tudor (1989) undertook a comprehensive content analysis of 990 horror films produced between 1931 and 1984. Horror films elicit fear by introducing a 'monstrous' threat into a stable situation. Tudor found that 'science' is historically the most frequent type of monstrous threat in horror films (251 out of 990, or 25 per cent). There has, however, been a broad decline in the proportion of science-based horror films after 1960. This decline does not necessarily indicate a change in public attitudes towards science, but it does represent a change in the production of horror films, as psychological horror took over as the dominant threat in the 1970s.

Peter Weingart and his colleagues undertook a quantitative study of 222 films of all genres, created over 80 years, looking for both recurring themes and changing patterns in the depiction of science in cinema (Weingart et al. 2003). Unsurprisingly, given its dominance in news media (Pellechia 1997), medical science is the most common research field depicted in films, followed by the physical sciences (chemistry and physics). These fields are also the most likely to be shown as 'ethically problematic' and to have scientist characters working in secret laboratories. In addition, Weingart et al. (2003) find that depictions of scientists are predominantly white, male and American. The overwhelming picture painted by both these studies is a cinematic history expressing deep-rooted fears of science and scientific research in the 20th century.

Cultural meanings

The studies discussed above fall under the category of 'traditional' or 'quantitative' content analysis. However, many researchers utilise a broader definition of content analysis that encompasses qualitative methods, including framing analysis (Altheide 1996). The most active area of research into the representation of science in cinema has been in what Jon Turney (1998) refers to as the 'cultural history of images', where textual analysis of fictional films provides researchers with a gauge of social concerns, social attitudes and social change regarding science and technology. Popular cultural products, like fictional films, not only reflect ideas about science and technology, they also construct perceptions for both the public and scientists in a mutual shaping of science and culture. The phrase 'movie scientist' does not often have positive connotations. For most people, the phrase conjures up images of Colin Clive as Dr Frankenstein maniacally repeating 'He's alive!' as his creature is brought to life in the classic 1931 Universal horror film *Frankenstein*. The mad scientist may be the most recognisable movie scientist, but this is not the only image of scientists on the screen. In her comprehensive study of representations of scientists in literature and film, Roslynn Haynes (1994) identifies six recurrent scientist stereotypes: the alchemist/mad scientist, the absent-minded professor, the inhuman rationalist, the heroic adventurer, the helpless scientist, and the social idealist. Depictions of scientists are particularly important as they represent the public face of science. Renato Schibeci and Libby Lee argue that cinematic images of scientists play a significant role in constructing students' 'science citizenship' by putting science in its socio-cultural context (Schibeci and Lee 2003).

These stereotypes recur in cinema because they possess narrative utility. Stereotypes are cinematic shorthand. Audiences easily recognise these scientist caricatures, so film-makers do not need to take up valuable screen time establishing character backgrounds. While these six basic stereotypes recur in cinema, they do not appear equally across genres. Horror films feature the mad scientist, comedies are the realm of the absent-minded professor and dramas predominantly feature social idealists. Likewise, action films incorporate heroic scientists, while science fiction films embrace inhuman rationalists and helpless scientists.

The prevalence of these cinematic stereotypes varies over time. The helpless scientist who loses control of his experiments was a common subject in films of the early 20th century, such as *Reversing Darwin's Theory* (1908). Although the stereotype of the scientist losing control of his experiments continues throughout cinema history, it takes on more ominous overtones as experiments have more dire consequences. The 1920s and 1930s, on the other hand, were the heyday of the mad scientist character, of which Clive's Dr Frankenstein is representative (Tudor 1989; Toumey 1996; Skal 1998; Frayling 2005). Unlike the helpless scientist, the character of the mad scientist has become so recognisable as a stereotype that the character now exists mainly in self-referential parodies such as *Young Frankenstein* (1974) and satires such as *Dr Strangelove* (1964).

The 1930s and 1940s were also the peak of the scientist biopic in Hollywood. Hollywood's approach to scientist biopics of this time period can be summed up by two words: miracle and tragedy (Elena 1997). *The Story of Louis Pasteur* (1936) exemplifies the standard biopic formula, and the film's success started the Hollywood scientist biopic trend that lasted until the mid-1940s. In the film, Pasteur has to overcome dogmatic scientific thinking and personal tragedies in order to bring the 'truth' of bacteriology to the public. As with most biopics of this period, *The Story of Louis Pasteur* not only presents science as a heroic endeavour, it also teaches audiences how science serves as a fact-producing activity (Crawford 1997). In addition, the narratives of scientist biopics, especially those of inventors such as Thomas Edison, link the work of science into the capitalist system by depicting science as the underlying source of mass production (Böhnke and Machura 2003).

The perceived motivations of scientists in the Manhattan Project fuelled a host of films in the 1950s featuring amoral, rationalist scientists who deny any responsibility for the consequences of their research (Shortland 1988; Jones 2001; Vieth 2001; Frayling 2005). *The Thing from Another World* (1951) exemplifies the depiction of the inhuman rationalist and the danger this represents to humanity. It is the scientists' insistence on studying the frozen alien body that creates the film's crisis situation. As one character claims, 'Knowledge is more important than life!' Absent-minded professors, as in *The Nutty Professor* (1963) and *The Absent-Minded Professor* (1961), joined inhuman scientists as scientist stereotypes who appeared regularly in films of this time period.

The 1990s and 2000s saw the ascendance of the heroic scientist stereotype in film. The popularity of the disaster film genre provided numerous opportunities to depict heroic scientists, as in *Dante's Peak* (1997) and *The Core* (2003) (King 2000). What is unique to this period is that many of the heroic scientist characters are women (Flicker 2003). Several studies have questioned historic gender representations in cinema, particularly with regard to primatology and the *Jurassic Park* (1993) film series (Haraway 1989; Briggs and Kelber-Kaye 2000; Franklin 2000; Kanner 2006). Jocelyn Steinke (2005) surveyed 74 science-based Hollywood films of the 1990s and found that 33 per cent (25 films) featured female scientists and engineers. Contrary to previous depictions, female scientist characters in the 1990s were realistic and did not always conform to traditional gender stereotypes. However, female scientists still corresponded to traditional notions of femininity in appearance and dress, and romance was a dominant theme in these films. In addition, female characters reinforced social and cultural assumptions about the role of women in science and engineering. They were single, most did not have children, and some permitted male colleagues to take credit for their work, as in the case of Ellie Arroway in *Contact* (1997).

Table 4.1 Summary of dominant scientist stereotypes, common themes and representative films over time

Time period	*Scientist stereotypes*	*Scientific fields*	*Representative films*
1900–10	Helpless scientists	Electricity X-rays Evolution	*X-Rays* (1897) *Reversing Darwin's Theory* (1908)
1911–20	Helpless scientists	Eugenics	*Damaged Goods* (1914) *The Regeneration of Margaret* (1916)
1921–30	Mad scientists	Glands Engineering	*A Blind Bargain* (1920) *Metropolis* (1926)
1931–40	Mad scientists Biopics	Medicine	*Frankenstein* (1931) *The Story of Louis Pasteur* (1936)
1941–50	Biopics	Medicine Psychology	*Shining Victory* (1941) *Madame Curie* (1944)
1951–60	Amoral scientists	Space science Nuclear science	*Destination Moon* (1950) *Them* (1954)
1961–70	Absent-minded professors	Space science	*The Nutty Professor* (1961) *2001: A Space Odyssey* (1969)
1971–80	Amoral scientists	Ecology	*Silent Running* (1971) *Soylent Green* (1973)
1981–90	Helpless scientists	Computer science	*War Games* (1983) *Robocop* (1987)
1991–present	Heroic scientists	Genetic engineering Astronomy	*Jurassic Park* (1993) *Deep Impact* (1998)

Scientific research fields in popular films

As with scientific stereotypes, the prevalence of specific scientific disciplines in cinema varies over time. Most of the scientific topics that dominate cinematic portrayals of science between 1900 and 1930 emerge from scientific discoveries made at the end of the 19th century. Louis Lumière patented his cinematograph in the same year (1895) that William Roentgen discovered X-rays. It did not take long for film-makers to exploit X-rays in films such as *X-Rays* (1897). Electricity also captured the minds of film-makers, and numerous films of the 1900s and 1910s incorporated electricity as a 'miracle' substance, as in *The Wonderful Electric Belt* (1907). By the 1920s, endocrinology became the science of choice for film-makers. The endocrinologist Dr Serge Voronoff became an international celebrity in 1919 for convincing wealthy, elderly men that implanted monkey glands would make them younger and give them the potency of 20-year-olds. Monkey glands soon became a staple plot element in several horror films of the early 1920s, including *A Blind Bargain* (1922). There were dramatic films dealing with scientific topics in this period, but these were propagandistic films about controversial science-based social issues such as eugenics (Pernick 1996). Pro- and anti-eugenics groups used popular film as a battleground between 1910 and the mid-1920s, with pro-eugenics films such as *Damaged Goods* (1914) and anti-eugenics films such as *The Regeneration of Margaret* (1916).

In the 1950s, science saw unprecedented growth in its activities and prestige with the rise of the military–industrial complex, and society viewed scientific progress as the means to lead post-war society towards a utopian existence. Despite science's overall increased visibility, it was a single event, the dropping of atomic bombs on Japan, that shaped the predominant portrayal of science in 1950s cinema (Weart 1988; Hendershot 1999; Shapiro 2002). Films of this period are about science as power, whether for good or evil. The opposition between science's destructive power and its progressive possibilities play out in various science fiction films of the period, including the trend-setting *Them!* (1954). In her seminal academic work on science fiction films of the 1950s, 'The imagination of disaster', Susan Sontag argues that science fiction films offer audiences pleasure in watching the 'aesthetics of destruction' while presenting morality plays about dangers inherent to science and technology (Sontag 1965: 102). For Sontag, science fiction films play to the medium's strength in visualisation, what she calls its 'sensuous elaboration', in communicating these dangers (ibid.: 101).

Nuclear science was not the only field heavily featured in films of the 1950s and 1960s. Starting with the ground-breaking *Destination Moon* (1950), space science became a major theme in cinema. Space films significantly shaped American space policy through their impact on American public opinion by showing space as an exciting and, most of all, technologically achievable adventure (McCurdy 1997). While radiation and space science dominated the 1950s and 1960s, popular cinema did address other aspects of scientific research, including the discovery of the double helical nature of DNA (Kirby 2003c) and advances in the human sciences (Vieth 2001).

Sparked by Rachel Carson's (1962) *Silent Spring*, films of the 1970s show an overriding concern with ecological disaster (Lambourne et al. 1990; Ingram 2000; Brereton 2005). There were a multitude of environmentally based science fiction, eco-horror and revenge-of-nature films in the early 1970s, including *Frogs* (1972) and *Soylent Green* (1973). Many of these films focus on issues of human over-population and resource use, and convey an impression that governmental ineptitude or inaction is to blame for these problems. By the 1980s and into the 1990s, the trend shifted to more serious dramas that moved the emphasis from governmental action towards corporate responsibility and individual responsibility in films such as *Silkwood* (1983) and *Erin Brockovich* (2000).

Computer science emerges as a strong theme in fictional films of the 1980s, as cinema grappled with two distinct aspects of our relationship to digital technologies. In the first instance, these films question the notion that humanity is really in control of our cybernetic creations, as in *War Games* (1983) and *The Terminator* (1984) (Glass 1989; Goldman 1989; Dinello 2006). Other films feature human-like robot/android/cyborg characters, including *Blade Runner* (1982) and *Robocop* (1987). Such artificially created humans in cinema represent the most effective way to gauge the range of definitions of humanness, as audiences must decide if these characters are actually 'human' (Telotte 1995; Wood 2002; Bukatman 2003). As Donna Haraway (1991) contends, 'cyborg bodies' show how the boundary between organisms and machines has eroded to the point of being invisible.

The 1990s and 2000s have been well studied with regard to cinema and its impact on the cultural meanings of genomics and genetic engineering. Although cinema is not the focus of their study, Dorothy Nelkin and M. Susan Lindee's ground-breaking work on genetics in popular culture demonstrates fiction's considerable input into shaping the 'cultural meaning' of DNA (Nelkin and Lindee 1995). Several films over the past 25 years have influenced the cultural meanings of genomics and genetic engineering, including *Boys From Brazil* (1978), *Twins* (1988) and *The Island of Dr Moreau* (1996) (Van Dijck 1998; Kirby 2002; Jörg 2003; Kirby and Gaither 2005). Two films, in particular, stand out for the quantity of scholarly attention they have received over the past 10 years: *Jurassic Park* and *Gattaca* (1997).

Jurassic Park's perceived impact on public perceptions of biotechnology explains a good deal of the scholarly commentary on the film. Cultural criticism of *Jurassic Park* has been diverse, with arguments being made about the metaphors of genetic engineering (Turner 2002), the film's simultaneous critique and exploitation of commercialism (Balides 2000), genetic engineering's transformation of the concept of 'life' (Franklin 2000), and the intersection of *Jurassic Park* with scientific culture (Stern 2004). In addition, scholars have discussed *Jurassic Park*'s influence on the BBC's *Walking with Dinosaurs* television series and on natural history film-making in general (Scott and White 2003).

While *Jurassic Park* is about the power of genetic engineering unwittingly to unleash monsters, *Gattaca* is about our power to shape humanity itself (Briggs and Kelber-Kaye 2000; Kirby 2000, 2004; Clayton 2002; Wood 2002). *Gattaca* is a rarity among film in its serious exploration of the bioethical issues surrounding human genetic manipulation, which explains why it has received so much critical examination. *Gattaca* contains the messages that we are more than the sum of our genes, and that being human means that we are able to 'transcend' our genetic obstacles. *Gattaca* is a unique film in that it does not question the morality of genomic intervention but, rather, asks the audience to consider what they are losing if they remove these genomic flaws. Ultimately, genetic engineering films of the 1990s and 2000s are about genetics as a science of information, control, transformation and identity.

Cinema, audience research and media effects

Although the difficulties and limitations of media effects research are well documented (Bryant and Zilmann 1994), several empirical studies of science in the media suggest that fictional representations can have an influence on public attitudes towards science. Most of these studies, however, examine science on television (Gerbner 1987; Sparks et al. 1997). As Nisbet et al. (2002) demonstrate, different media affect perceptions differently, so it is not certain how science in cinema relates to these studies. Despite a widespread belief that cinema has negative impacts both on science literacy and on public perceptions of science, the effects on the public of science in cinema remain a relatively sparse area of research. There have been a few traditional media-effects studies, mostly revolving around *The Day After Tomorrow* (2004), but most studies on this question are sociologically based, and a growing area of research examines the impact fictional films have on science itself.

Science literacy and science education

The National Science Foundation (NSF) has singled out fictional media as a corrosive influence on science literacy and the public's critical thinking skills. According to the NSF's *Science and Engineering Indicators – 2006* (National Science Board 2006), 'forms of popular culture, such as books and movies, affect what people know about science and shape their attitudes toward science-related issues'. While this viewpoint seems to be widely accepted within the scientific community, there is very little empirical research to back up this claim. Despite this lack of evidence, scientists and scientific organisations are active in addressing scientific inaccuracies in Hollywood cinema, usually by releasing 'real science of . . .' reviews. A 'real science of' analysis consists of a scientist, or scientists, critiquing a fictional text in terms of what they see as inaccurate scientific content. For example, astronomer Stephen Maran (1998) discusses the 'real science' in the fictional films *Armageddon* (1998) and *Deep Impact* (1998). The National Institutes of Health maintain a long-running public film series involving scientists' critiques of science in films in order to combat cinema's supposed negative influence on science literacy. There is also a growing utilisation of cinema as a pedagological tool for teaching science (Dubeck et al. 1994; Edwards 1997; Brake and Thornton 2003; Rose 2003). The idea is that the use of cinema can attract young people to science courses because of its popularity, and cinema's visual nature can be used to grab their attention while they are in the classroom. There is even a degree programme at the University of Glamorgan in Wales, UK, 'Science and Science Fiction', designed around the use of science fiction in the classroom to teach science.

Science propaganda and entertainment education

A field of practitioner-based inquiry is the intentional use of fiction to promote social agendas known as 'environmental education' (Singhal and Rogers 2002). These works are almost exclusively related to using television as a means for changing individual behaviour regarding public health issues. It may well be that television is a more effective medium for entertainment education than cinema, because audiences feel they 'know' characters they encounter on a weekly basis. Cinematic cautionary tales about public health issues were numerous in the early 20th century, and many were made with the cooperation of public health officials, physicians and medical researchers (Pernick 1978, 1996; Lederer and Parascandola 1998; Fedunkiw 2003; Boon 2005). Likewise, in the 1930s the US Government and other concerned organisations used films to promote technological development (Kline 1997).

When scientists and scientific organisations are involved in production, they can help craft narrative in popular films to promote their research fields, their scientific institutions, and their own scientific work (Kirby 2003a). While not necessarily seeking to promote specific causes, many scientific organisations view cinema as a means of promoting their work. Large, prominent research institutions have even set up divisions devoted to seeking out relationships with the entertainment industry. NASA, for example, established its Entertainment Industry Liaison in the late 1960s, and still actively seeks out collaboration with Hollywood (NASA 2002). It has recently been intricately involved in the production of numerous films, including *Deep Impact*, *Mission to Mars* (2000) and *Space Cowboys* (2000). For each of these films, NASA provided technical advice on the sets, access to its scientists for scientific advice, script analysis, and the use of facilities and equipment. NASA even authorised the use of its logo by a fictional text for the first time in *Mission to Mars* and *Space Cowboys*. Even scientific organisations that do not chase Hollywood, such as the National Severe Storms Laboratory [*Twister*] and the US Geological Survey [*Dante's Peak*], are not shy about promoting their 'brand' in cinema if approached by film-makers.

Audience reception studies

Before and after the release of the recent Hollywood blockbuster *The Day After Tomorrow*, survey- and focus group-based studies of public attitudes about global warming were conducted in Germany (Reusswig et al. 2004), Britain (Lowe et al. 2006) and the USA (Leiserowitz 2004). While there was no evidence for a significant shift in overall public opinion in the USA, 50 per cent of those who saw the film indicated that they were 'somewhat more or much more worried' about global warming, whereas only 1 per cent said that they became 'less worried'. Surprisingly, in Germany, a significant number of respondents indicated that their conviction that global warming was real had weakened after they viewed the movie. Katie Mandes from the Pew Center on Global Climate Change suggests that most Europeans had a stable notion of the effects of climate change before they went to see the film, and that the movie's version of events, global warming leading to global cooling, contradicted their expectations (quoted by Schiermeier 2004). British viewers were significantly more concerned about climate change, and some respondents expressed stronger motivation to act on climate change after seeing the film. All these studies agree, however, that the film raised awareness of the issue, whether or not it actually changed attitudes.

Impact on science

The most exciting research into science, cinema and media effects involves an examination of cinematic representation's impact on science itself. The cinematic apparatus emerged out of the scientific research of Eadweard Muybridge and Etienne-Jules Marey, who were looking for technological means of studying animal movement in the late 19th century. Since that time, moving pictures have remained an integral part of scientific research (Cartwright 1995; Gaycken 2002; Ostherr 2005; Landecker 2006). But cinema has impacts on science beyond being a research tool. Fictional films can influence science by enhancing funding opportunities, promoting research agendas, influencing public controversies, and playing a role in intra-specialist communication. Films can have significant impact on policy debates as politicians and members of the general public often use fictional stories to frame their concerns about science and technology (Mulkay 1996; Huxford 2000; Nerlich et al. 2001). Lily Kay (2000) offers the useful notion of the 'technoscientific imaginary' to account for shared representational practices both within science and in the broader culture. Technoscientific imaginary encompasses all the narratives, both scientific and public, that frame an issue and give it its cultural value. Kay's conceptions of the technoscientific imaginary compliments the work of Serge Moscovici (1976) on social representation theory, in that cinematic representations provide society with shared sets of representations that allow social discussion, even if those discussions reveal divergent differences. Whether or not it has any real impact on the public, a film's assumed impact on public opinion can give it utility within the political arena, as was the case for *The Day After Tomorrow*, because it provided shared social representations and became part of the technoscientific imaginary (Nisbet 2004).

Two films in 1998, *Armageddon* and *Deep Impact*, became an integral part of the scientific and political rhetoric surrounding near-Earth object protection (Davis 2001; Kirby 2003b; Mellor 2007). Scientists helped construct these fictional narratives as powerful rhetorical devices that allow audiences to 'virtually witness' a conception of nature where near-Earth objects are an imminent threat. These films, along with other media texts, constructed a technoscientific imaginary that was equally useful for the disparate scientific communities of civilian and military scientists, as well as various political communities. Science in cinema, however, rarely exists as a solitary entity. One need only look at *Jurassic Park* in its incarnations as a novel, film, comic book, computer game, television documentaries and news articles to see the high degree of intertextuality in science-based media. The interplay between popular texts and formal scientific discourse is what Heather Schell (1997) calls 'genre interpenetration'. Genre interpenetration was clearly evident in the case of *Outbreak* (1995) as popular science texts, documentaries, political treatises and scientific works all borrowed imagery and narratives from the film. News media incorporated the film's images and narratives in their coverage of a real-life outbreak of Ebola in Zaire that occurred while the film was in cinemas in the USA and Europe (Vasterman 1995; Semmler 1998; Ostherr 2005). Cinema fits into Bruce Lewenstein's (1995) 'web' model of science communication, in that cinema, other mass media and technical media interact in complex ways, informing and referring to each other.

Conclusions

To be successful, a modern science-based film must adhere to a sense of scientific authenticity. Yet the production of a Hollywood film is a complicated process, where multiple actors may have competing agendas for science. Film-makers take a flexible approach in determining what 'authenticity' means in the context of fictional cinema. Thus scientific accuracy will always take a backseat to storytelling. The point of movies is not to devise 'accurate/educational' communications about science, but to produce images of science that are entertaining. Understanding the work of scientists in Hollywood is important, but science consultants are only a small component of the production process. There is still a need for scholars to uncover exactly how, and why, film-makers produce filmic images of science. How do scriptwriters approach science? How important is science for special effects technicians? What role does science play for production designers or the art department? Truly to appreciate cinema as a mode of science communication and its role in the public understanding of science, it is crucial that we understand fully science's place in the production process.

As indicated by the focus of the NSF's bi-annual *Science and Engineering Indicators* on science literacy, the deficit model for science communication still enjoys favour among scientists and scientific organisations. The time has come to think about science and cinema outside its impact, if any, on science literacy. Those interested in cinema and scientific literacy need to follow Schibeci and Lee's (2003) example by embracing Alan Irwin's (1995) concept of science citizenship. Cinematic depictions of science involve the production and presentation of an image of science, whether or not the science has anything to do with 'real science'. Cinematic images carry a cultural currency that both reflects and influences public attitudes towards the scientific enterprise. Scientists and scientific organisations concerned about science literacy, like the NSF, need to recognise the value of scholarly work on the cultural meanings of science with respect to cinematic science. Preliminary studies suggest how cinema interacts with other media and scientific discourse to create a technoscientific imaginary that has impacts on the systems of science. Filmic images can have an impact on the public's conceptions of science by provoking reactions, from encouraging excitement to instilling fear about science and technology, and often both. The question remains – exactly what, and how strong, is this impact?

While there is significant hand-wringing at the NSF, National Institutes of Health, Wellcome Institute and other funding agencies over science in cinema, there is no sense of urgency, and there are very few resources available for studying the problem. One of the impediments to studying science in cinema, why few scholars make this a full-time area of study, is a long-standing perception of movies as inconsequential 'fluff'. Science in cinema is at once a corrosive influence on science literacy and a source of insignificant depictions that audiences recognise as a fantasy world. These contradictory perceptions result in a patchwork research landscape, where scholars from various fields often dip their hands into the 'science and cinema' pool before moving back to their disciplinary homes to pursue more 'serious' research topics. The work outlined in this chapter shows an emerging picture of a complicated research field where cinema science can have a major impact on our concept of science communication and the public understanding of science. The time has come to move away from merely indicating concern about science in movies and to start treating this topic as a significant and unified field of academic research.

Notes

1 My discussion in this chapter is predominantly confined to popular, fictional films. While there is a good deal of similarity between cinema and television as visual media, television has its own production practices, marketing, dissemination routes, sites of reception and cultural contexts, and science on television has its own body of scholarly literature. The same issues and lessons can be applied to other fictional media, including literature. This study focuses exclusively on mainstream Hollywood cinema – while there are numerous studies of European and world cinema, space constraints here necessitated a focus on one region, the choice of American films also being justified by the significant impact of American cinema on world culture.

2 It is not possible to give complete descriptions of films within this chapter. For more information about the films discussed here, visit the Internet Movie Database (www.imdb.com).

Suggested further reading

Haynes, R. (1994) *From Faust to Strangelove*, Baltimore, MD: Johns Hopkins University Press.
Lambourne, R., Shallis, M. and Shortland, M. (1990) *Close Encounters?*, New York: Adam Hilger.
McCurdy, H. (1997) *Space and the American Imagination*, Washington, DC: Smithsonian.
Nelkin, D. and Lindee, S. M. (1995) *The DNA Mystique*, New York: W.H. Freeman.
Tudor, A. (1989) *Monsters and Mad Scientists*, Oxford: Basil Blackwell.
Vieth, E. (2001) *Screening Science*, Lanham, MD: Scarecrow.
Wood, A. (2002) *Technoscience in Contemporary American Film*, Vancouver, BC: University of British Columbia Press.

Other references

Altheide, D. (1996) *Qualitative Media Analysis*, London: Sage.
Balides, C. (2000) 'Jurassic post-Fordism: tall tales of economics in the theme park', *Screen*, 41: 139–60.
Brereton, P. (2005) *Hollywood Utopia*, Bristol: Intellect Books.
Böhnke, M. and Machura, S. (2003) '*Young Tom Edison – Edison, the Man*: biopic of the dynamic entrepreneur', *Public Understanding of Science*, 12: 319–33.
Boon, T. (2005) 'Health education films in Britain, 1919–1939: production, genres and audiences', in Harper, G. and Moor, A. (eds) *Signs of Life*, London: Wallflower, 45–57.
Bousé, D. (2000) *Wildlife Films*, Philadelphia, PA: University of Pennsylvania Press.
Brake, M. and Thornton, R. (2003) 'Science fiction in the classroom', *Physics Education*, 38: 1–4.
Briggs, L. and Kelber-Kaye, J. I. (2000) 'There is no unauthorized breeding in Jurassic Park: gender and the uses of genetics', *NWSA*, 12: 92–113.
Bryant, J. and Zillmann, D. (eds) (1994) *Media Effects*, Hillsdale, NJ: Lawrence Erlbaum.
Bukatman, S. (2003) *Matters of Gravity*, Durham, NC: Duke University Press.
Carson, R. ([1962] 1987) *Silent Spring*, Boston, MA: Houghton Mifflin.
Cartwright, L. (1995) *Screening the Body*, Minneapolis, MN: University of Minnesota Press.
Clayton, J. (2002) 'Genome time', in Newman, K., Clayton, J. and Hirsch, M. (eds) *Time and the Literary*, London: Routledge, 31–59.
Crawford, T. (1997) 'Screening science: pedagogy and practice in William Dieterle's film biographies of scientists', *Common Knowledge*, 6: 52–68.
Davis, D. (2001) '"A hundred million hydrogen bombs": total war in the fossil record', *Configurations*, 9: 461–508.
Dinello, D. (2006) *Technophobia!* Austin, TX: University of Texas Press.
Dubeck, L. W., Moshier, S. E. and Boss, J. E. (1994) *Fantastic Voyages*, New York: American Institute of Physics.
Edwards, R. (1997) 'Evolutionary biology at the movies', *Journal of College Science Teaching*, March: 333–38.
Elena, A. (1997) 'Skirts in the lab: Madame Curie and the image of the woman scientist in the feature film', *Public Understanding of Science*, 6: 269–78.
Fedunkiw, M. (2003) 'Malaria films: motion pictures as a public health tool', *American Journal of Public Health*, 93: 1046–56
Flicker, E. (2003) 'Between brains and breasts: women scientists in fiction film: on the marginalization and sexualization of scientific competence', *Public Understanding of Science*, 12: 307–18.
Frank, S. (2003) 'Reel reality: science consultants in Hollywood', *Science as Culture*, 12: 427–69.
Franklin, S. (2000) 'Life itself: global nature and genetic imaginary', in Franklin, S., Lury, C. and Stacey, J. (eds) *Global Nature, Global Culture*, London: Sage, 188–227.

Frayling, C. (2005) *Mad, Bad and Dangerous*, London: Reaktion.
Gaycken, O. (2002) 'A drama unites them in a fight to the death: some remarks on the flourishing of a cinema of scientific vernacularization in France, 1909–1914', *Historical Journal of Film, Radio and Television*, 22: 353–74.
Gerbner, G. (1987) 'Science on television: how it affects public conceptions', *Issues in Science and Technology*, 3: 109–15.
Glass, F. (1989) 'The "New Bad Future": Robocop and 1980s' sci-fi films', *Science as Culture*, 5: 6–49.
Goldman, S. (1989) 'Images of technology in popular films: discussion and filmography', *Science, Technology and Human Values*, 14: 275–301.
Haraway, D. (1989) *Primate Visions*, London: Routledge.
—— (1991) *Simians, Cyborgs and Women*, London: Routledge.
Haynes, R. (1994) *From Faust to Strangelove*, Baltimore, MD: Johns Hopkins University Press.
Hendershot, C. (1999) *Paranoia, The Bomb, and 1950s Science Fiction Films*, Bowling Green, OH: Bowling Green State University Press.
Huxford, J. (2000) 'Framing the future: science fiction frames and the press coverage of cloning', *Journal of Media and Cultural Studies*, 14: 187–99.
Ingram, D. (2000) *Green Screen*, Exeter: University of Exeter Press.
Irwin. A. (1995) *Citizen Science*, London: Routledge.
Irwin, A. and Wynne, B. (eds) (1996) *Misunderstanding Science?* Cambridge: Cambridge University Press.
Jones, R. (2001) '"Why can't you scientists leave things alone?": science questioned in British films of the post-war period (1945–1970)', *Public Understanding of Science*, 10: 1–18.
Jörg, D. (2003) 'The Good, the Bad, the Ugly: Dr Moreau goes to Hollywood', *Public Understanding of Science*, 12: 297–305.
Kanner, M. (2006) 'Going on instinct: gendering primatology in film', *Journal of Popular Film and Television*, 33: 206–12.
Kay, L. (2000) *Who Wrote the Book of Life?*, Stanford, CA: Stanford University Press.
King, G. (2000) *Spectacular Narratives*, London: I.B. Tauris.
Kirby, D. (2000) 'The new eugenics in cinema: genetic determinism and gene therapy in *GATTACA*', *Science Fiction Studies*, 27: 193–215.
—— (2002) 'Are we not men?: the horror of eugenics in *The Island of Dr Moreau*', *Paradoxa*, 17: 93–108.
—— (2003a) 'Scientists on the set: science consultants and communication of science in visual fiction', *Public Understanding of Science*, 12: 261–78.
—— (2003b) 'Science consultants, fictional films and scientific practice', *Social Studies of Science*, 33: 231–68.
—— (2003c) 'The threat of materialism in the age of genetics: DNA at the drive-in', in Rhodes, G. (ed.) *Horror at the Drive-In: Essays in Popular Americana*, Jefferson, NC: McFarland, 241–58.
—— (2004) 'Extrapolating race in *Gattaca*: genetic passing, identity, the new eugenics, and the science of race', *Literature and Medicine*, 23: 184–200
Kirby, D. and Gaither, L. (2005) 'Genetic coming of age: genomics, enhancement, and identity in film', *New Literary History*, 36: 263–82.
Kline, R. (1997) 'Ideology and the new deal "fact film" *Power and the Land*', *Public Understanding of Science*, 6: 19–30.
Lambourne, R., Shallis, M. and Shortland, M. (1990) *Close Encounters?*, New York: Adam Hilger.
Landecker, H. (2006) 'Microcinematography and the history of science and film', *Isis*, 97: 121–32.
Lederer, S. and Parascandola, J. (1998) 'Screening syphilis: Dr Ehrlich's magic bullet meets the public health service', *Journal of the History of Medicine*, 53: 345–70.
Leiserowitz, A. (2004) 'Before and after *The Day After Tomorrow*: A U.S. study of climate change risk perception', *Environment*, 46: 22–37.

Lewenstein, B. (1995) 'From fax to facts: communication in the cold fusion saga', *Social Studies of Science*, 25: 403–36.
Lowe, T., Brown, K., Dessai, S., Doria, M., Haynes, K. and Vincent, K. (2006) 'Does tomorrow ever come? Disaster narrative and public perceptions of climate change', *Public Understanding of Science*, 15: 435–57.
Maran, S. (1998) 'Movie myths vs scientific reality', *Washington Post*, 12 August.
McCurdy, H. (1997) *Space and the American Imagination*, Washington, DC: Smithsonian.
Mellor, F. (2007) 'Colliding worlds: asteroid research and the legitimization of war in space', *Social Studies of Science*, 37: 499–531
Mitman, G. (1999) *Reel Nature*, Cambridge, MA: Harvard University Press.
Moscovici, S. (1976) *Social Influence and Social Change*, London: Academic Press.
Mulkay, M. (1996) 'Frankenstein and the debate over embryo research', *Science, Technology, and Human Values*, 21: 157–76.
NASA (2002) 'NASA explores future collaborations with Hollywood', *NASA Headquarters Bulletin*, October: 7.
National Science Board (2006) *Science and Engineering Indicators – 2006*, Arlington, VA: National Science Foundation.
Nelkin, D. and Lindee, S. M. (1995) *The DNA Mystique*, New York: W.H. Freeman.
Nerlich, B., Clarke, D. D. and Dingwall, R. (2001) 'Fictions, fantasies and fears: the literary foundations of the cloning debate', *Journal of Literary Semantics* 30: 37–52.
Nisbet, M. (2004) 'Evaluating the impact of *The Day After Tomorrow*: can a blockbuster film shape the public's understanding of a science controversy?' *Skeptical Inquirer*, 16 June, www.csicop.org/scienceandmedia/blockbuster
Nisbet, M. C., Scheufele, D. A., Shanahan, J., Moy, P., Brossard, D. and Lewenstein, B. V. (2002) 'Knowledge, reservations, or promise? A media effects model for public perceptions of science and technology', *Communication Research*, 29: 584–608.
Ostherr, K. (2005) *Cinematic Prophylaxis*, Durham, NC: Duke University Press.
Pellechia, M. (1997) 'Trends in science coverage: a content analysis of three US newspapers', *Public Understanding of Science*, 6: 49–68.
Pernick, M. (1978) 'Thomas Edison's tuberculosis films: mass media and health propaganda', *Hastings Center Report*, 8: 21–27.
—— (1996) *The Black Stork*, Oxford: Oxford University Press..
Peters, H. (1995) 'Interaction of journalists and scientists', *Media, Culture and Society*, 17: 31–48.
Reingold, N. (1985) 'Metro-Goldwyn-Mayer meets the atom bomb', in Shinn, T. and Whitley, R. (eds) *Expository Science*, Dordrecht: D. Reidel, 229–45.
Reusswig, F., Schwarzkopf, J., Pohlenz, P. (2004) 'Double impact: the climate blockbuster, the day after tomorrow and its impact on the German cinema public', *PIK Report*, 92. www.pik-potsdam.de/news-1/press-releases/archive/2004/pr92.pdf
Rose, C. (2003) 'How to teach biology using the movie science of cloning people, resurrecting the dead, and combining flies and humans', *Public Understanding of Science*, 12: 289–96.
Schell, H. (1997) 'Outburst! A chilling true story about emerging-virus narratives and pandemic social change', *Configurations*, 5: 93–133.
Schibeci, R. and Lee, L. (2003) 'Portrayals of science and scientists, and "science for citizenship"', *Research in Science and Technological Education*, 21: 177–92.
Schiermeier, Q. (2004) 'Disaster movie highlights transatlantic divide', *Nature*, 431: 4.
Scott, K. and White, A. (2003) 'Unnatural history? Deconstructing the *Walking with Dinosaurs* phenomenon', *Media, Culture and Society*, 25: 315–32.
Semmler, I. (1998) 'Ebola goes pop: the Filovirus from literature into film', *Literature and Medicine*, 17: 149–74.
Shapiro, J. (2002) *Atomic Bomb Cinema*, London: Routledge.

Shortland, M. (1988) 'Mad scientists and regular guys: images of the expert in Hollywood films of the 1950s', *Proceedings of the Joint Meeting of the British Society for History of Science and the History of Science Society*, Manchester, UK, July 1988, 291–98.

Singhal, A. and Rogers, E. (2002) 'A theoretical agenda for entertainment education', *Communication Theory*, 12: 117–35.

Skal, D. (1998) *Screams of Reason*, New York: Norton.

Sontag, S. ([1965] 2004) 'The imagination of disaster', reprinted in Rickman, G. (ed.) *The Science Fiction Film Reader*, New York: Proscenium, 98–113.

Sparks, G. G., Nelson, C. L. and Campbell, R. G. (1997) 'The relationship between exposure to televised messages about paranormal phenomena and paranormal beliefs', *Journal of Broadcasting and Electronic Media*, 41: 345–59.

Steinke, J. (2005) 'Cultural representations of gender and science: portrayals of female scientists and engineers in popular films', *Science Communication*, 27: 27–63.

Stern, M. (2004) '*Jurassic Park* and the moveable feast of science', *Science as Culture*, 13: 347–72.

Telotte, J. (1995) *Replications*, Chicago, IL: University of Illinois Press.

Toumey, C. (1996) *Conjuring Science*, New Brunswick, NJ: Rutgers University Press.

Tudor, A. (1989) *Monsters and Mad Scientists*, Oxford: Basil Blackwell.

Turner, S. (2002) 'Jurassic Park technology in the bioinformatics economy: how cloning narratives negotiate the telos of DNA', *American Literature*, 74: 887–909.

Turney, J. (1998) *Frankenstein's Footsteps*. New Haven, CT: Yale University Press.

Van Dijck, J. (1998) *Imagenation*. London: Macmillan.

Vasterman, P. (1995) 'The Hollywood plague', *Albion Monitor*, 19 August.

Vieth, E. (2001) *Screening Science*, Lanham, MD: Scarecrow.

Weart, S. (1988) *Nuclear Fear*, Cambridge, MA: Harvard University Press.

Weingart, P. and Pansegrau, P. (2003) 'Introduction: perception and representation of science in literature and fiction film', *Public Understanding of Science*, 12: 227–28.

Weingart, P. with Muhl, C. and Pansegrau, P. (2003) 'Of power maniacs and unethical geniuses: science and scientists in fiction film', *Public Understanding of Science*, 12: 279–87.

Wood, A. (2002) *Technoscience in Contemporary American Film*, Vancouver, BC: University of British Columbia Press.

Yandetti, M. (1978) 'Atomic scientists and Hollywood: the beginning or the end?' *Film and History*, 8: 73–88.

5

Of deficits, deviations and dialogues

Theories of public communication of science

Massimiano Bucchi

Theoretical and empirical research in public communication of science has a relatively short history compared with the long-standing practice of communicating science to the public. It was only in 1992, for instance, that a dedicated scholarly journal, *Public Understanding of Science*, was founded. This chapter seeks to contribute to the theoretical understanding of science communication by, first, outlining the key elements of the traditional conception, still implicitly or explicitly widespread within science communication practice and policy, of public communication of science. I then review some of the studies that have challenged different facets of this conception. Finally, I raise the question of which alternative models can best help us understand the contemporary interactions between scientific knowledge and the general public.

The traditional conception of public communication of science

Scientific communication addressed to the layman has a long tradition. Consider the numerous popular science books written in the 18th century to satisfy growing public interest, especially among women, including Algarotti's *Newtonianism for Ladies* or de Lalande's *L'Astronomie des Dames*, the numerous accounts of scientific discoveries published in the daily press, or the great exhibitions and fairs that showed visitors the latest marvels of science and technology (Raichvarg and Jacques 1991).

However, communication practices in science have developed mainly in relation to two broad processes: the institutionalisation of research as a profession with higher social status and increasing specialisation; and the growth and spread of the mass media.

The idea that science is 'too complicated' for the general public to understand became established particularly as a result of advances made in physics during the early decades of the 1900s. In December 1919, when observations made by astronomers during a solar eclipse confirmed Einstein's general theory of relativity, the *New York Times* gave much prominence to a remark attributed to Einstein himself: 'At most, only a dozen people in the world can understand my theory' (cited by Pais 1982: 309).

This idea underpins a widespread conception, if not an outright 'ideology', of the public communication of science. Other cornerstones of the conception are the need for mediation between scientists and the general public, made necessary by the complexity of scientific notions; the singling out of a category of professionals and institutions to perform this mediation (science journalists and, more generally, popularisers of science, museums and science centres); and use of the metaphor of translation to describe this mediation.

This 'diffusionist' conception, unquestionably simplistic and idealised, which holds that scientific facts need only be transported from a specialist context to a popular one, is rooted in the professional ideologies of two of the categories of actors involved. It legitimates the social and professional role of the 'mediators' – popularisers, particularly science journalists – who comprise the most visible and most closely studied component of the mediation. It also authorises scientists to proclaim themselves extraneous to the process of public communication so that they may be free to criticise errors and excesses – especially in terms of distortion and sensationalism. There has thus arisen a view of the media as a 'dirty mirror' held up to science, an opaque lens unable adequately to reflect and filter scientific facts.

In addition, this vision has emphasised the public's inability to understand and appreciate the achievements of science due to prejudicial public hostility as well as to misrepresentation by the mass media, and adopts a linear, pedagogical and paternalistic view of communication to argue that the quantity and quality of the public communication of science should be improved. In order to recover this deficit, public and private bodies – especially since the mid-1980s – have launched schemes aimed at promoting public interest in and awareness of science. These initiatives have included 'open days', now a routine feature of many laboratories and research institutions, science festivals and training courses in science journalism.

In summary, the traditional, diffusionist conception of public communication of science incorporates a notion of:

1. the media as a channel designed to convey scientific notions, but often unable to perform this task satisfactorily due to lack of competences and/or predominance of other priorities (e.g. commercial interests);
2. the public as passive, whose default ignorance and hostility to science can be counteracted by appropriate injection of science communication;
3. science communication as a linear, one-way process in which the source context (specialist elaboration) and target context (popular discourse) can be sharply separated, only the former influencing the latter;
4. communication as a broader process concerned with the transfer of knowledge from one subject or group of subjects to another;
5. knowledge as being transferable without significant alterations from one context to another, so that it is possible to take an idea or result from the scientific community and bring it to the general public.

Although such notions have mutually reinforced each other, and have sometimes overlapped to some extent, it could be noted that one of the labels most frequently used to refer to this whole constellation of notions, the 'deficit' model, refers in particular to the second assumption above. Regarding the first assumption, within the past three decades research on media coverage of science issues has gone beyond the mere stereotype of the diffusionist conception and its pleas for greater accuracy, for closer interaction between journalists and specialist sources, and, in general, for efforts to minimise the elements that cause 'disturbance' in communication between scientists and the general public, which otherwise would be straightforward. In this light, for instance, the role of newsmaking routines and journalistic priorities in shaping science coverage has been articulated (see Chapter 2 in this volume). Likewise, the media's criteria for selecting 'scientific experts' to comment on a specific issue do not necessarily coincide with those of the scientific community. Journalists – particularly news journalists as opposed to 'specialised' science journalists – have also sometimes reacted forcefully to the expectation that their criteria should correspond to those of scientists, seeing it as their professional duty to express public concerns and demands, and describing their mission in terms of public opinion's need for information, thus justifying their indifference to the priorities of the scientific agenda (Hansen 1992; Peters 1995).

However, long-period analysis of the treatment of scientific themes by the non-specialist press shows that it presents scientific activity as largely 'progressive', as beneficial to society, and as consensual. Such coverage is found to adhere closely to specialist sources, often cited directly or indirectly, and in linguistic terms is often not particularly distant from specialist communication.[1]

Is the public scientifically illiterate?

The diffusionist (pedagogical–paternalistic) conception of the communication of science has long informed studies on public scientific knowledge. First conducted in the USA during the 1950s, research on the general public's interest in and awareness of science and scientific information has, since the 1980s, become common in numerous countries. The results of this research have frequently been used to decry the public's scant interest in science and its excessively low level of 'scientific literacy', and to call for quantitative and qualitative improvements in science communication addressed to the public at large. Since the early 1990s, these assumptions have been strongly criticised on several grounds. It has been pointed out that the equation between public understanding and the ability to answer questions about science has long restricted the discussion to the somewhat tautological observation that members of the public do not reason in the same way as professional scientists. Also disputed are the assumed links between exposure to science in the media, level of knowledge, and a favourable attitude toward research and its applications. As regards biotechnologies, for example, recent research has shown a substantial degree of scepticism and suspicion, even among the sections of the population most exposed to scientific communication and best informed about biotechnological topics (Bucchi and Neresini 2002). In general, it does not seem that the opposition of certain sectors of the general public to particular technical–scientific innovations is due solely to the presence of an information deficit. Rather, the phenomenon requires more systematic and detailed analysis.

More generally, the disjunction between expert and lay knowledge cannot be reduced to a mere information gap between experts and the general public as envisaged by the deficit model. Lay knowledge is not an impoverished or quantitatively inferior version of expert knowledge; it is qualitatively different. Factual information is only one ingredient of lay knowledge, in which it interweaves with other elements (value judgements, trust in the scientific institutions, the person's perception of his or her ability to put scientific knowledge to practical use) to form a corpus no less sophisticated than specialist expertise (Wynne 1989, 1995).

The role of scientists

And what about scientists? Are they truly extraneous to these processes, passively at the mercy of the discursive practices of journalists and the incomprehension of the public? Studies on the public communication of science tell us that they are not: around 80 per cent of French researchers report that they have had some experience of popularising science through the mass media, and similar conclusions were reached in a study on US scientists by Dunwoody and Scott (1982). Almost one-fifth of the articles on science and medicine published in the past 50 years by the Italian daily newspaper *Corriere della Sera* have been written by science/medical experts (Bucchi and Mazzolini 2003). According to a broad survey of British scientists and journalists, already in the early 1990s more than 25 per cent of the articles on science that appeared in the press started from initiatives – press releases, announcements of discoveries, interviews – by researchers and their institutions, a percentage that is likely to have increased since then (Hansen 1992). Researchers are often among the most assiduous users of science coverage by the media, on which they draw to select from the enormous mass of publications and research studies in circulation. A paper published in the prestigious *New England Journal of Medicine* is three times more likely to be cited in the scientific literature if it has first been mentioned by the *New York Times* (Phillips 1991). The visibility of scientists in the media tends to display a pyramid structure very similar to that of the distribution of other resources and remunerations in the scientific community. At the top of the pyramid stand a very small number of 'celebrities' who are frequently consulted on non-scientific issues as well – Nobel prize-winners being a typical example – and below them, a broad base with very sporadic visibility (Goodell 1977). These results have also prompted sociologists of science to interest themselves in the public communication of science, a topic on which their contribution has long been marginal in comparison with other disciplines including social psychology, linguistics and media studies. This lack of interest in the public presentation and awareness of science can be explained by considering sociologists of science to be the most sophisticated victims of the traditional conception. As long as the public communication of science was considered a practice detached from science, it was of scant relevance to those interested in the influence of social factors on scientific activity.

Public communication of science as the continuation of scientific debate by other means

Science studies are highly critical of the traditional conception of the public communication of science. Instead of the sharp distinction between science and its popularisation, they propose a 'continuity' model of scientific communication (Cloître and Shinn 1985; Hilgartner 1990). Along the continuum thus envisaged, gradual differences can be discerned in the diverse contexts and styles of communication/reception that exist in the exposition of scientific ideas. Cloître and Shinn (1985) identify the following four main stages in the process of scientific communication.

- *Intraspecialist level* – the most distinctively esoteric level, typified by papers published in specialised scientific journals. Empirical data, references to experimental work and graphs predominate.
- *Interspecialist level* – includes various kinds of texts, from interdisciplinary articles published in 'bridge journals' such as *Nature* and *Science* to papers given at meetings of researchers belonging to the same discipline, but working in different areas.
- *Pedagogic level* – described by Fleck (1935) as 'textbook science', where the theoretical corpus is already developed and consolidated and the current paradigm is presented as complete. The emphasis is on the historical perspective and the cumulative nature of the scientific endeavour.
- *Popular level* – covers for instance articles on science published in the daily press and the 'amateur science' of television documentaries. Cloître and Shinn point to the quantity of metaphorical images in these texts and their marked attention to issues concerning health, technology and the economy.

A typology of this kind, which presents science communication as a continuity of texts with differences in degree, not in kind, across levels, invites us to imagine a sort of trajectory for scientific ideas that leads from the intraspecialist expository context to the popular one, passing through the intermediate levels. This is a trajectory congruent with theories, from Fleck's to Latour's, on the construction of scientific facts. We may take as an example the tortuous process studied by Fleck, which led from a vague popular idea of 'syphilitic blood' to introduction of the Wassermann reaction and definition of the clinical distinctiveness of syphilis. This highly provisional definition, hedged about by doubts and methodological caveats, rapidly became an incontrovertible certainty in the eyes of the general public. Fleck used this example to reflect on the path followed by a medical–scientific notion from what he called the esoteric circle (the specialist community) to the exoteric one (the general public). Fleck compared a report on a clinical examination drawn up by one specialist for another with a report prepared for a general practitioner, which 'does not represent the knowledge of the expert. It is vivid, simplified and apodictic' (Fleck 1935, Eng. tr. 1979: 113).

Specialist exposition – the 'science of the journals' – is provisional and tentative. But when a theory makes its entry into the textbooks, it partly loses these features and is presented to the reader as generally accepted by the medical–scientific community:

in other words, it becomes a 'fact'. A further step comes with the exposition characteristic of popular science; here 'the fact becomes incarnated as an immediately perceptible object of reality' (ibid. 125). At the popular level, doubts and disclaimers disappear: the distinctions and nuances of specialist knowledge condense into elementary and compact formulas: AIDS is HIV; psychoanalysis studies 'complexes'; the neurological theory that hypothesises a division of tasks between the two hemispheres of the brain is transformed into a sharp antithesis between 'right-dominated' and 'left-dominated' people. The communicative path from specialist to popular science can thus be illustrated as like a funnel that removes subtleties and shades of meaning from the knowledge that passes through it, reducing it to simple facts attributed with certainty and incontrovertibility (Figure 5.1). Fleck stresses that this progressive solidification of knowledge then exerts an influence on specialists themselves. 'Certainty, simplicity, vividness originate in popular knowledge. That is where the expert obtains his faith in this triad as the ideal of knowledge. Owing to simplification, vividness and absolute certainty [popular knowledge] appears secure, more rounded and more firmly joined together' (ibid. 113, 115). The passage of a scientific notion through these various levels therefore cannot be described as the simple translation of an object from one communicative context to another. Each step – and this is one of the central messages of Fleck's book – involves a change in the notion. By way of analogy, something similar happens to characters and stories in literature. For example, none of Arthur Conan Doyle's original works contains the expression 'Elementary, my dear Watson'. Only after its introduction in a theatre production of the detective's adventures did the phrase come to epitomise Sherlock Holmes in the popular imagination.

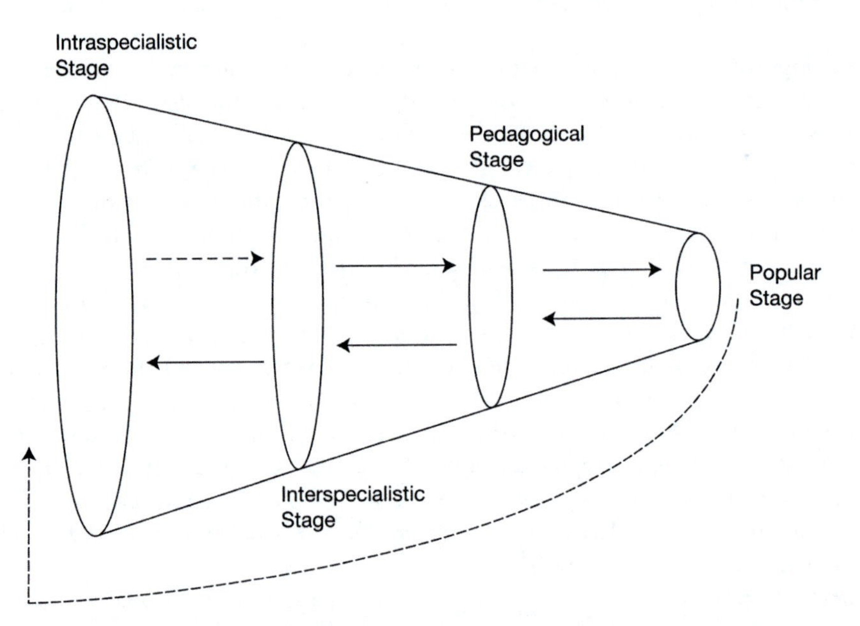

Figure 5.1 A model of science communication as a continuum.
Source: Bucchi 1998

Taking this intuition to its extreme, studies by sociologists of science based on the continuity model consider the level of popular communication to be the final (and often decisive) stage in the process of stylisation, 'distancing from the research front', and production of factuality and incontrovertible truth that constructs scientific evidence (Collins 1987). As Whitley (1985: 13) puts it, 'The more removed the context of research is from the context of reception in terms of language, intellectual prestige and skill levels, the easier it is to present their work as certain, decontextualised from the conditions of its production, and authoritative.'

The continuity model can be considered a useful frame of reference insofar as it describes some sort of ideal flow of communication in routine circumstances. However, in some cases, public communication seems able to perform a more sophisticated role, for example in bringing a scientific matter to attention of policy-makers and, consequently, of the wider scientific community, as happened in the case of sickle-cell anaemia in the USA (Balmer 1990).

In this case we can speak of a 'deviation' to the public level, because the discourse did not follow the usual trajectory, but passed directly at the public level to then influence specialist circles. The importance of appealing to the public in particular cases of change of controversy or paradigm has been variously hypothesised and studied (Jacobi 1987). For example, the wide and enthusiastic coverage given in 1919 by the daily press to the solar eclipse observations as confirming Einstein's theory of relativity – *The Times*' headline was 'Revolution in Science: New Theory of the Universe: Newtonian Ideas Overthrown' – played a crucial role in publicly settling an issue that was still being debated within specialist circles (Gregory and Miller 1998).

Mention has already been made of how scientists make use of the information and images that circulate at the public level. Cloître and Shinn (1985) document how specialists appropriated a metaphor (the ant in the labyrinth) originally used by popular science texts to explain the Brownian motion of particles. Around one-third of the scholars involved in the debate on the connection between mass extinction of dinosaurs and a meteor collision with the Earth – another controversy with broad public resonance – stated that they had heard of the impact hypothesis from the mass media (Clemens 1994).

It has been argued that scientific discourse at the public level may, in some cases, resemble certain forms of political discourse in that it is only apparently public. It is not really addressed to the public, but instead is intended to reach a large number of colleagues rapidly. To do so, it uses the public level as a shared arena where it is not necessary to comply with the constraints of specialist communication. This prerogative of the public level is particularly important when communication must pass through several disciplinary sectors (a case in point being the hypothesis on the extinction of the dinosaurs, which concerned palaeontologists, geologists and statisticians) or several categories of actors. In the late 19th century, Pasteur's struggle to legitimise the anthrax vaccine and, more generally, the idea that diseases could be prevented by appropriate inoculation with the infectious agent, meant that physiologists, doctors, veterinarians and farmers had to be addressed simultaneously. This difficult task was achieved by means of a public experiment organised in 1881 on a farm, where vaccinated and non-vaccinated cattle were infected with anthrax before the eyes of hundreds of people, including French and foreign newspaper reporters, who wrote numerous detailed articles on Pasteur's success. Communication at the public level enabled the French physiologist to play down still unclear theoretical issues by emphasising practical ones (of great importance to some groups in his audience, e.g. farmers and politicians) such as the effectiveness and cheapness of his method. Moreover, immunisation and the related practice of inoculation had long been familiar to the lay peasant culture (Bucchi 1997).

In 1919, Einstein was able simultaneously to address different disciplinary audiences (physicists, astronomers, mathematicians) through the popular press by giving interviews and writing articles on his theory of relativity (Gregory and Miller 1998). The scientists who argued that the depletion of the ozone layer was due to CFC found the widely publicised image of the ozone 'hole' to be an effective device through which to alert researchers, politicians, environmentalists and public opinion to the emergency. The rapid public consensus achieved with the Montreal Protocol of 1987, which provided for international agreements to reduce the CFC emissions responsible for ozone depletion, indirectly reinforced the status of a body of knowledge that was still being carefully debated by specialists (Grundmann and Cavaillé 2000).

When a new sector of research is being established or consolidated – as happened with climate studies or the neurosciences in past decades – the public arena is vital if researchers are to communicate among different disciplines. Communicating in public enables scientists not only to talk – albeit indirectly – among themselves (as Fleck pointed out), but also to gain recognition and construct a shared identity in terms of research interests and methods, thereby laying the basis for institutionalisation of their sector.

In cases of 'deviation', therefore, the science communication process should be depicted as much more complex. In these situations, the public discourse of science does not receive simply what is filtered through previous levels, but may instead find itself at the centre of the dynamics of scientific production. By and large, when talking about the public communication of science, we are referring to at least two different things:

- a routine trajectory, consensual and non-problematic, which is adequately described by the continuity model – despite its ideological connotations, 'popularisation' is a sufficiently appropriate term for this process;
- an alternative trajectory, which is the one represented by deviation to the public level, so that public communication acquires even greater salience and a more articulated role compared with specialist debate.

There are major formal and substantial differences between these two trajectories. At a formal level, when the popularisation mode is activated, scientific problems are more frequently addressed in settings devoted explicitly to the communication of science: popular science magazines and the science pages of newspapers. Placing scientific notions in these media 'frames' gives them legitimacy and enhances their credibility. The most obvious example is the museum medium: the display of a scientific artefact in a museum tends automatically to confer the status of incontrovertible 'fact' upon it (Macdonald and Silverstone 1992).

When deviation occurs, scientific problems appear more frequently in generic media settings as well, such as the news sections of newspapers and television newscasts. The scientific facts, as well as the networks of professional and institutional actors surrounding them, may be consolidated, as the continuity model envisages, but they may also be dissolved, deconstructed or simply manipulated by social groups for their own purposes. The funnel does not necessarily taper off; it may expand again towards the specialist levels. In these situations, social actors outside the research community, such as activists or representatives of patients' associations may play a significant role in the definition of scientific facts, as in the case of research on AIDS (Grmek 1989; Epstein 1996).

Study of public scientific discourse in cases of deviation enables account to be taken of the 'plurality of the sites for the making and reproduction of scientific knowledge' (Cooter and Pumfrey 1994: 254), and also gives a more sophisticated role to the public, who the funnel model tends to reduce to no more than a passive source of external support. A theory or a scientific finding may consequently enjoy different status and robustness at different levels of communication. Thus the Big Bang may represent *the* explanation of the origin of the universe in the popular domain, despite the doubts and distinctions expressed in the specialist one.

While deviation may be an opportunity to evade the rules and constraints of the popularisation process, it is often regarded with suspicion by the scientific community. When scientific problems are pushed into the public arena, they lose some of the status that they may still enjoy in such popularisation frames as the scientific journals or the science sections of newspapers. They may, for example, be subject to problem concatenation processes, or undergo life cycles like all other issues of public interest. Moreover, they can presumably also be manipulated and introduced into the public arena by actors external to the scientific community, such as journalists, policy-makers or the leaders of movements and associations.

This helps explain the growing efforts by scientists to extend their control over communication with the public. Scientific institutions organise seminars on these matters and invite journalists to 'live laboratory life' for brief periods; researchers write booklets advising their colleagues on how to handle the media. Research institutes now make much use of public relations offices and similar devices, not to exclude the possibility of deviations (which would be difficult to achieve), but to extend the scientific community's control over recognition of 'crises' and over the activation of deviation processes so that the latter can be put to *ad hoc* use or criticised. Scientists often engage in deviation (public communication as part of the process by which a scientific fact is produced), but camouflage it as popularisation (the diffusion of scientific knowledge with pedagogic intent). Many of the misunderstandings that surround the debate on the public communication of science probably arise because popularisation expectations are attributed to communications that, in reality, perform deviation functions – they serve to regulate the scientific debate for 'internal' purposes – and *vice versa*.[2] There is tension within the scientific community between the institutionalisation of deviation – its absorption into ordinary expository practice (popularisation) in order to prevent its 'uncontrolled abuse' – and its defence as a sort of 'emergency exit' for certain situations, and as a potential source of scientific change and innovation.

Can knowledge be transferred?

If the public and the media have been problematised since the initial reflections on science communication – and science has been more recently problematised in the same context (see e.g. Wynne 1995; Michael 2002) – communication itself as a concept has, so far, rarely been problematised. Much of the diffusionist ideology of science communication fundamentally rests on a notion of communication as transfer. For at least 60 years such a notion has been the dominant paradigm for describing communication – for scholars, practitioners and lay persons – as a process concerned with the transfer of knowledge from one subject or group of subjects to another subject or group of subjects. The widespread and unquestioned use of keywords such as 'reception', 'flow', 'distortions' and 'target' when discussing communication is indicative of the power and pervasiveness of this transfer metaphor. Within this notion, successful communication is defined as the achieved transfer of information from one party to another; for instance, a public communication initiative in the area of genetics could be considered successful if a fraction of the knowledge available to the scientific community on this topic is acquired by a certain target public. This notion takes for granted, among other things, the possibility of transferring knowledge without significant alteration from one context to another, so that we can simply take an idea from the scientific community and bring it to the general public; and that the same knowledge in different contexts will result in the same attitudes and eventually in the same type of behaviour. Beginning in the 1950s, a great number of studies in the area of communication – and in particular, mass communication – have challenged some of the core elements of the transfer vision. Studies have shown, among other things, that different types of filter can contribute to make the transfer a selective process. Filters include selective perception of media messages, previous motivations and attitudes of audiences, and communication intermediaries such as opinion leaders.[3] In the specific domain of science communication, several empirical and theoretical contributions have critically addressed the idea of transfer during the past two decades.

A necessarily selective list of the aspects which have been pointed out includes:

- the non-linearity of the communication process; science communication need not necessarily spring from specialised contexts, but can also originate in popular, non-specialised arenas (Lewenstein 1995a, 1995b; Bucchi 1996, 1998);
- the reception of science communication is not a passive process, but a complex set of active transformative processes that can, in turn, have an impact on the core scientific debate itself (Wynne 1989, 1995; Epstein 1996);
- specialist exposition of science theories and results (the source of transfer in the traditional paradigm) cannot be sharply separated from popular exposition (the target of transfer), despite the fact that distinctions between the two forms of exposition are often used by scientific actors as a rhetorical strategy (Hilgartner 1990);
- the science communication process can be better represented as a continuous sequence of expository levels, gradually shifting one into another with differences in degree and not in kind, mutually influencing one another (Cloître and Shinn 1985; Hilgartner 1990; Lewenstein 1995a; Bucchi 1996, 1998).

There are also several indications that public discourse about many science issues has not arisen as a filtered or trickled-down version of specialist discourse. In his study of genetics in popular culture, to continue with this example, Jon Turney (1998) has shown that key achievements in terms of research agenda, including Watson and Crick's discovery of DNA structure, did not receive immediate attention by the general media; on the other hand, popular ideas on the transformation of species and modification of man had a much longer history, as documented for instance by the famous claim by French novelist Emile Zola – 30 years before the rediscovery of Mendel's laws of heredity – that 'heredity has its laws, just like gravitation' (Zola 1871; see also Lewontin 1996).

Understanding of science communication may benefit from stepping out of the transfer metaphor to investigate the multiple interactions of specialist and popular discourse. Communication may thus be seen as intense short-circuiting or cross-talking between those discourses – rather than as plain transfer – taking place under certain circumstances and centring on key discursive 'boundary objects' (e.g. gene, DNA, Big Bang, AIDS) lying at the intersection between specialist and popular levels.

Such objects make communication possible without necessarily requiring consensus, for an object may be interpreted and used in quite different ways within different types of discourses. 'Gene' could thus be seen as a boundary object, a label employed in both specialist and public contexts and thereby providing a common language, although translated in different ways in a laboratory conversation and in a car advertisement.[4] In this light, the spell intrinsically tying communication to understanding, as in the deficit vision, can finally be broken.

A model of science communication as cross-talk also implies seeing communication not simply as a cause – for instance, of changes in opinions and attitudes among the public, due to the transfer of certain results or ideas – but also as the result of developments in both discourses, allowing the formation of an intersection zone. It is reasonable to hypothesise that, once formed, this intersection would facilitate exchanges across different discourses, reinforcing itself in a recursive fashion. Another advantage could be seen in the model's recapturing a view of communication as a process – which sustains (and has to be sustained by) actors' interaction – rather than as a taken-for-granted point of departure.

From deficit to dialogue, from dialogue to participation – and beyond?

During the past decade, enduring public concern over certain science and technology issues despite significant communication efforts; growing citizen demand for involvement in such issues; and multiplying examples of non-experts actively contributing to shaping the agenda of research in fields such as biomedicine have led to a rethinking of the very meaning of public communication of science in several arenas. For instance, in 2000 a report from the UK House of Lords acknowledged the limits of science communication based on a paternalistic, top-down science–public relationship, and detected a 'new mood for dialogue'. In 2002, the Committee on the Public Understanding of Science (Copus), set up in 1985 by the

Royal Society and other institutions to support public awareness activities, was also brought to an end by its very founders, who reached the conclusion that 'the top-down approach which Copus currently exemplifies is no longer appropriate to the wider agenda that the science communication is now addressing'. In many countries, and at the European level, funding schemes and policy documents shifted their keywords from 'public awareness of science' to 'citizen engagement'; from 'communication' to 'dialogue'; from 'science *and* society' to 'science *in* society'. Initiatives have flowered that are aimed at eliciting public input on science and technology issues, and decision-making about science and technology. A notion of 'knowledge co-production' has been introduced by scholars to describe intense forms of participation of non-experts in the definition and accreditation of scientific knowledge – as when patients' organisations actively contribute to defining the priorities of medical research, or when citizens' groups gather epidemiological data that lead experts to rethink the cause of a certain pathology (Brown and Mikkelsen 1990). These forms have been interpreted as representing a major change not only with regard to the deficit model, but also with regard to its sociological critiques. According to Callon (1999), for instance, the critical version of public understanding of science – as reflected in the dialogic option – shifts the priority from 'the education of a scientifically illiterate public' to the need and right of the public to participate in the discussion, on the assumption that 'lay people have knowledge and competencies which enhance and complete those of scientists and specialists'. However, both models are seen as sharing 'a common obsession: that of demarcation. [The first model], in a forceful way, and [the second model], in a gentler, more pragmatic way, deny lay people any competence for participating in the production of the only knowledge of any value: that which warrants the term "scientific"' (ibid. 89). On this basis, the need has been invoked for another, more substantial shift to a model of knowledge co-production in which non-experts and their local knowledge can be conceived as neither an obstacle to be overcome by virtue of appropriate education initiatives (as in the deficit model), nor an additional element that simply enriches professionals' expertise (as in the critical–dialogical model), but rather as essential for the production of knowledge itself. Expert and lay knowledge are not produced independently in separate contexts to encounter each other later; rather, they result from common processes carried forward in 'hybrid forums' in which specialists and non-specialists can interact (Callon et al. 2001).

Does the change of keywords actually reflect a change in the practice and understanding of science communication? Or is it – as some scholars have suggested – in many cases a reappearance of the traditional, deficit model in a new guise (Stilgoe et al. 2005; Trench 2006)? How are these changes redefining, if ever, the role of science communication? Which theoretical model(s) can best help us interpret this changing scenario? Or, to cite another scholar, 'how dead is the deficit model?' (Trench 2006).

To answer these questions, I suggest that we first need to pay attention to the issue of context. One of the lessons from the 'sociological turn' of science communication studies is that public communication of science cannot be understood in a vacuum; rather, it should always be viewed not only in the context of expert/citizen interactions, but also in the broader context of science in society.

This apparently simple recommendation has several significant implications. One is that we cannot straightforwardly apply models of science communication (such as a diffusionist, popularisation notion of science communication), largely developed within the context of a science performed by relatively few state-based institutions, to a science characterised by pervasive relationships with the markets, global outlook and a strong public relations push (for which scholars have coined the label 'PUS inc.'; Bauer and Gregory 2007). Moreover, contemporary science is increasingly challenging the very notion of a sharp distinction between producers and users of knowledge, which rests at the basis of a diffusionist, deficit, transfer vision of science communication.[5] Companies, environmental organisations and patients' groups have established themselves as legitimate sources and providers of science communication.

A feature of the contemporary science in society context is also its intrinsic heterogeneity and fragmentation: communication is subject to the contradictory pressures of knowledge privatisation and commodification, open access and sharing of research results, and citizens' demands for greater involvement. All this makes implausible the use of a single science communication model to account for the varieties of contemporary expert/public configuration.

Table 5.1 sets out three key models of expert/public interaction – deficit,[6] dialogue and participation[7] – together with their vision of communication and their broader ideological contexts. These models should be conceived as ideal types, rather than as mutually exclusive categories. Most communicative situations would have to be described by a combination of the three models. In this framework, the deficit model does not need to disappear: it becomes the default, 'zero degree' of expert/public interaction processes. This is why it is important to distinguish the many different facets of such a model. While there are strong cases for dropping its expectation that public scepticism can be overcome by injecting knowledge, its top-down, transfer vision of communication may be a reasonable proxy to describe situations characterised, for instance, by a

Table 5.1 A multi-model framework of science communication (adapted from Trench 2006)

Communication model	*Emphasis*	*Dominant versions in science communication*	*Aims*	*Ideological contexts*
Transfer Popularisation One-way, one-time	Content	Deficit	Transferring knowledge	Scientism Technocracy Rhetoric of the knowledge economy
Consultation Negotiation Two-way, iterative	Context	Dialogue	Discussing implications of research	Social responsibility Culture
Knowledge co-production, deviation Multi-directional, open-ended	Content and context	Participation	Setting the aims, shaping the agenda of research	Civic science Democracy

low degree of public mobilisation, on science issues that have relatively low public resonance.[8] Over time, public/expert interaction with regard to a certain issue may move across models and their combinations: for instance, an emerging topic such as nanotechnology may lend itself to deficit-like communication in its initial stages, and later become the subject of public consultation/mobilisation; knowledge produced on a rare genetic pathology in situations of intense interaction between experts and non-experts may subsequently become the focus of a deficit-like communication initiative. Studies highlighting the connection between increase in the public salience of a certain science issue – or even in the level of knowledge – and mounting concern on the part of the public (Mazur 1981; Bucchi and Neresini 2002) might have actually grasped the 'tip of the iceberg' of these shifting configurations.

Coherence between communication patterns and the aims and ideological contexts deserves particular attention, as it may also help to clarify why institutions such as the European Commission have encountered difficulties in matching their claims for public participation in science and technology. A participatory, co-production approach to science communication appears difficult to couple with the emphasis on technocracy and rhetoric of the knowledge economy that form the basis of much EU policy strategy in the area of research, and rather lends itself to more traditional, deficit-transfer communication strategies (Trench 2006). Unlike deficit configurations, participation is also, by definition, multidirectional, open-ended and potentially subject to conflict. Some degree of apprehension for this open-endedness may be regarded as a key factor accounting for the sometimes resurgent temptation, on the part of research bodies and other institutions, to 'tame' unruly public participation through formal initiatives, or bluntly preaching dialogue and participation while practising the deficit. More generally, a continuous tension exists between opening up the black box of deficit communication for participation and, instead, putting participation back into the deficit box, with groups and institutions publicly struggling to impose their communicative definition of the situation – deficit, dialogue or participatory-like. A meta-level of science communication can be imagined, in which actors are constantly engaged to define (in participatory, dialogic or deficit form) the configuration of their interaction on a certain issue.

A communication pattern should also not necessarily be overlapped with the aims and interests of a specific category of actors. Research and policy institutions may (in a deficit-like communicative fashion) promote dialogic/participatory situations; citizens may contribute (in a dialogic/participatory fashion) to relegate into the deficit realm an issue on which they have little interest in participating, or on which they feel comfortable reducing their role to quasi-passive spectators of knowledge as channelled by experts for their own cultural benefit, aesthetic appreciation or entertainment.

In this light, rather than 'which model of science communication accounts best' for expert–public interactions, one of the key sociological questions becomes 'under what conditions do different forms of public communication of science emerge?'

While a detailed analysis of such conditions would require a treatment of its own, a tentative list could, in principle, include:

- the degree of public salience of a certain science issue;
- the level of public mobilisation on that and neighbouring issues;

- the visibility and credibility of science institutions and actors involved;
- the degree of controversy/disagreement among science experts, as perceived by the public;
- the degree of institutionalisation and the stability of professional boundaries in the science field of concern;
- the degree of social consensus on the overarching political and cultural context of science issues.

It may be expected, for instance, that an issue in the field of particle physics, with low public impact and mobilisation, little controversy among experts, propelled by visible research institutions, in a context in which understanding of the fundamental laws of nature is a socially shared and undisputed aim, may lend itself to a deficit-like pattern in which the public is invited and willing to appreciate the spectacle of science's achievements. Likewise, an issue such as genetically modified organisms, touching many publicly relevant themes including food, safety, biodiversity and resource distribution, with a certain amount of experts' disagreement as publicly perceived, propelled by corporate actors in a context highly sensitive, alerted and mobilised to questions of environment and globalisation, was unlikely to be containable in the deficit box. However, variations in the above-mentioned – and other potential – conditions may be reflected in a significant redefinition of the communication pattern. If an astrophysics result is framed as 'the Holy Grail of cosmology', as happened with the discovery in 1992 of radioactivity in the outer reaches of the known universe, taken to represent the echo of the Big Bang at the origin of the universe, the situation may slide into a more dialogic, open communication pattern in which the very boundaries between science and religion may be at stake (Miller 1994; Bucchi 2000).

It should be emphasised that the social, political and cultural contexts have a bearing not only on the introduction of new knowledge by the experts. Emerging trends in popular discourse can give a completely different status and meaning to already existing scientific results, turning a transfer-deficit situation into an intense communicative short-circuiting. Despite a significant advance in human cloning announced by a team of scientists in 1993, cloning was not an issue in countries such as Italy until the announcement of Dolly established a connection to a debate that had developed over issues such as embryos, in vitro fertilisation and abortion (Neresini 2000).

The broader political context may also be decisive in setting the scene for communicative interaction. Switzerland's or Scandinavia's tradition of civic participation is reflected in the relevance given to that participation with regard to science, to the point of being incorporated into legislative prescription and dedicated institutional agencies (e.g. Joss and Bellucci 2002).

Some general historical trends can be identified in the variations of these conditions. For instance, it is hard to deny that the increasing level of general education among citizens of many countries, or the expanded potential access to science information through the internet, have made participatory configurations more frequent and accessible today, particularly in areas such as biomedicine and the environment (Nowotny et al. 2001; Chapter 13 in this volume). Other broad trends may include the increasingly pervasive role of the media in questioning not only policy decisions on science but more specifically the connection between expertise and policy making; and the rising demand for public participation as part of more general criticism of the capacity of traditional democracies to represent and include citizens' points of view when addressing global challenges, with crucial decisions being more and more taken at levels not directly subject to citizens' influence – the so called 'democratic deficit' that is frequently a matter of concern with regard to, for instance, European or international institutions (Andersen and Burns 1996; Levidow and Marris 2001).

Other conditions may be much less stable. Several studies in sociological and historical perspective suggest, for instance, that the inclination of scientists to open up their communicative boundaries to non-experts is not a new, nor a steadily rising phenomenon, but could rather be described in terms of alternating cycles of openness and closure (deviation and popularisation) in a sort of pendular movement (Hirschman 1982). The consequences of these conditions seem far from straightforward.

For instance, when researchers mobilise in the public arena to protest against budget cuts or against state regulation of certain research fields, or simply advocate greater public concern with science, they may contribute to a growing public perception of scientific expertise as interest-laden, thus damaging the credibility of traditional decision-making arrangements that involve only experts and policy-makers (Bucchi and Neresini 2004). This, in turn, suggests an ironic and somewhat paradoxical generalisation of the above-mentioned 'open-endedness': citizens' pressures for more participation, which have contributed to undermining the deficit approach, may have been stimulated, among other things, by scientists' advocacy of that self-same approach.

On this view, one should also resist the temptation to interpret the different analytical models of interactions among experts and the public as a chronological sequence of stages in which the emerging forms obscure the previous ones, with the dialogue version obliterating the deficit one, or the participatory version substituting for the dialogue one. The interpretive framework proposed here seeks to account for the simultaneous coexistence of different patterns of communication that may coalesce, depending on specific conditions and on the issues at stake.

Communication should not be reified as a circumscribed, static event, nor as a prerogative that can be switched on and off at will. Rather, it should be viewed as a process that fluidly assumes different contingent configurations. A certain notion of the relationship between professional experts and the public – for instance, as segregated categories in the deficit model, or as inextricably intertwined as in the co-production model – is in itself a result of, and not a precondition for, the struggles, negotiations and alliances taking place in those configurations.

This theoretical framework certainly does not provide the science communicator with instructions as simple and appealing as those offered by the transfer/deficit approach; no easy 'switch' – to borrow another term from contemporary genetics research – to press (e.g. 'more communication!', 'focus the target!' 'clarify the message!') in order to produce the desired outcome among the public. Nor does it support the expectation that the aims of the 'old' deficit approach may finally be fulfilled by upgrading to the next 'communication fix', be it called dialogue or participation. The present framework eventually makes the process of public communication of science – and thereby the activities in which science communication practitioners are routinely engaged – more relevant, not only as a means to achieve certain objectives, but also as a central space in which to understand (and participate in) the interacting transformations of both science and public discourse. In this perspective, communication is not simply a technical tool functioning within a certain ideology of science and its role in economic development and social progress, but has to be recognised as one of the key dynamics at the core of those co-evolutionary processes (Nowotny et al. 2001; Jasanoff 2004, 2005), redefining the meanings of science and the public, knowledge and citizenship, expertise and democracy.

Notes

1 Cf. Lewenstein (1995a); Bucchi and Mazzolini (2003); Stephens (2005). Casadei (1994), for example, has conducted comparative lexical analyses of popular science texts, textbooks and specialist articles on physics, finding similar levels of technicality in the three genres, with the maximum level not in the specialist texts but in the textbooks.
2 To draw another analogy, deviation with respect to popularisation can be considered equivalent to a scientific revolution with respect to normal science (Kuhn 1962).
3 The pervasive influence of the transfer metaphor has been seen in relation to its being, to some extent, embedded in languages such as English – lacking expressions to describe communication processes other than those related to 'transfer' and 'transportation' of messages (Reddy 1979).
4 In this sense, boundary objects are not so different from what the actor-network theory identifies as 'obligatory passage points' in translating interests and enrolling supporters for a scientific claim (Latour 1987), or what Moscovici (1961) locates at the heart of a social representation (its 'zero degree').
5 'Post-academic science' and 'mode-2 science' are some of the labels used by scholars to indicate these emerging configurations of research in contemporary society (Gibbons et al. 1994; Ziman 2000; Nowotny et al. 2001)
6 As has been noted, 'deficit' refers to a specific element of the model, the emphasis on the knowledge asymmetry between experts and the public as a basis and rationale for the communicative interaction. It would be more accurate to refer to this model as a 'diffusionist' conception that, beyond the deficit element, incorporates a notion of communication as unproblematic one-way transfer, having no impact on the processes of knowledge production (popularisation). However, as 'deficit' has become the standard label for the whole constellation among policy-makers and scholars, I use it hereafter with the same general meaning.
7 Public participation in science may be defined broadly as 'the diversified set of situations and activities, more or less spontaneous, organised and structured, whereby non-experts become involved, and provide their own input to, agenda setting, decision-making, policy forming and knowledge production processes regarding science' (Callon et al. 2001; Rowe and Frewer 2005; Bucchi and Neresini 2007). Under such a broad definition, participation encompasses not only formal participatory initiatives promoted by a certain sponsor (such as consensus conferences), but also a broad range of situations including public protests, referendum voting and patient initiatives. When seen from the point of view of experts/insiders, and of the consequences for specialist discourse, a participation pattern of interaction can be said to incorporate a deviation element.
8 Different reappraisals of the deficit model are given by Sturgis and Allum (2004); Dickson (2005).

Suggested further reading

Bucchi, M. (1998) *Science and the Media. Alternative Routes in Scientific Communication*, London and New York: Routledge.

Bucchi, M. and Neresini, F. (2007) 'Science and public participation', in Hackett, E. et al. (eds) *Science and Technology Studies Handbook*, Cambridge, MA: MIT Press, 955–1001.

Cooter, R. and Pumfrey, S. (1994) 'Science in popular culture', *History of Science*, 32: 237–67.

Fleck, L. (1935) *Entstehung und Entwicklung einer wissenschaftliche Tatsache* (Eng. tr. *Genesis and Development of a Scientific Fact*, Chicago, IL: University of Chicago Press, 1979).

Shinn, T. and Whitley, R. (1985) (eds) *Expository Science. Forms and Functions of Popularization*, Dordrecht: Reidel.

Other references

Andersen, S. and Burns, T. (1996) 'The European Union and the erosion of parliamentary democracy: a study of post-parliamentary governance', in Andersen, S. and Eliassen, K. A. (eds) *European Union – How Democratic is It?*, London: Sage, 227–51.

Balmer, B. (1990) 'Scientism, science and scientists', research paper, Science Policy Research Unit, University of Sussex, UK.

Bauer, M. and Gregory, J. (2007) 'From journalism to corporate communication in post-war Britain', in Bucchi, M. and Bauer, M. (eds) *Journalism, Science and Society: Science Communication between News and Public Relations*, London: Routledge: 33–52.

Brown, P. and Mikkelsen, E. (1990) *No Safe Place: Toxic Waste, Leukemia, and Community Action*, Berkeley, CA: University of California Press.

Bucchi, M. 1996 'When scientists turn to the public: alternative routes in science communication', *Public Understanding of Science*, 5: 375–94.

—— (1997) 'The public science of Louis Pasteur: the experiment on anthrax vaccine in the popular press of the time', *History and Philosophy of the Life Sciences*, 19: 181–209.

—— (1998) *Science and the Media. Alternative Routes in Scientific Communication*, London and New York: Routledge.

—— (2000) 'A public explosion: Big Bang in the UK daily press', in Dierckes, M. and von Grote, C. (eds) *Between Understanding and Trust: The Public, Science and Technology*, Reading: Harwood.

Bucchi, M. and Mazzolini, R. G. (2003) 'Big science, little news: science coverage in the Italian daily press, 1946–1997', *Public Understanding of Science*, 12: 7–24.

Bucchi, M. and Neresini, F. (2002) 'Biotech remains unloved by the more informed', *Nature*, 416: 261.

—— (2004) 'Why are people hostile to biotechnologies?' *Science*, 304: 1749.

—— (2007) 'Science and public participation', in Hackett, E. et al. (eds) *Science and Technology Studies Handbook*, Cambridge, MA: MIT Press, 955–1001.

Callon, M. (1999) 'The role of lay people in the production and dissemination of scientific knowledge', *Science, Technology & Society*, 4: 81–94.

Callon, M., Lascoumes, P. and Barthe, Y. (2001) *Agir dans un monde incertain: Essai sur la démocratie technique*, Paris: Seuil.

Casadei, F. (1994) 'Il lessico nelle strategie di presentazione dell'informazione scientifica: il caso della fisica', in De Mauro, T. (ed.) *Studi sul trattamento linguistico dell'informazione scientifica*, Roma: Bulzoni, 47–69.

Clemens, E. (1994) 'The impact hypothesis and popular science: conditions and consequences of interdisciplinary debate', in Glen, W. (ed.) *The Mass-Extinction Debates: How Science Works in a Crisis*, Stanford, CA: Stanford University Press.

Cloître, M. and Shinn, T. (1985) 'Expository practice: social, cognitive and epistemological linkages', in Shinn, T. and Whitley, R. (eds) *Expository Science. Forms and Functions of Popularization*, Dordrecht: Reidel, 31–60.

Collins, H. M. (1987) 'Certainty and the public understanding of science: science on television', *Social Studies of Science*, 17: 689–713.

Cooter, R. and Pumfrey, S. (1994) 'Science in popular culture', *History of Science*, 32: 237–67.

Dickson, D. (2005), 'The case for a "deficit model" of science communication', SciDev.Net, 28 June 2005.

Dunwoody, S. and Scott, B. (1982) 'Scientists as mass media sources', *Journalism Quarterly*, 59: 52–9.

Epstein, S. (1996) *Impure Science: AIDS, Activism and the Politics of Knowledge*, Berkeley, CA: University of California Press.

Fleck, L. (1935) *Entstehung und Entwicklung einer wissenschaftliche Tatsache* (Eng. tr. *Genesis and Development of a Scientific Fact*, Chicago, IL: University of Chicago Press, 1979).

Goodell, R. (1977) *The Visible Scientists*, Boston, MA: Little Brown.

Gibbons, M., Limoges, C., Nowotny, H., Schwartzman, S., Scott, P. and Trow, M. (1994) *The New Production of Knowledge: Dynamics of Science and Research in Contemporary Societies*, London: Sage.

Gregory, J and Miller, S. (1998) *Science in Public. Communication, Culture, and Credibility*, London: Plenum.

Grmek, M. D. (1989) *Histoire du SIDA*, Paris, Payot.

Grundmann, R. and Cavaillé, J. P. (2000) 'Simplicity in science and its publics', *Science as Culture*, 9: 353–89.

Hansen, A. (1992) 'Journalistic practices and science reporting in the British press', *Public Understanding of Science*, 3: 111–34.

Hilgartner, S. (1990) 'The dominant view of popularization', *Social Studies of Science*, 20: 519–39.

Hirschman, A. (1982) *Shifting Involvements. Private Interest and Public Action*, Princeton, NJ: Princeton University Press.

Jacobi, D. (1987) *Textes et Images de la Vulgarisation Scientifique*, Bern: Peter Lang.

Jasanoff, S. (2004) *States of Knowledge: The Co-production of Science and Social Order*, London: Routledge.

—— (2005) *Designs on Nature. Science and Democracy in Europe and the United States*, Princeton, NJ: Princeton University Press.

Joss, S. and Bellucci, S. (eds) (2002) *Participatory Technology Assessment: European Perspectives*, London: Centre for the Study of Democracy.

Kuhn, T. S. (1962) *The Structure of Scientific Revolutions*, Chicago, IL: Chicago University Press (2nd edn 1969).

Latour, B. (1987) *Science in Action*, Cambridge, MA: Harvard University Press.

Levidow, L. and Marris, C. (2001) 'Science and governance in Europe: lessons from the case of agricultural biotechnology', *Science and Public Policy*, 28: 345–60.

Lewenstein, B. (1995a) 'Science and the media', in Jasanoff, S. et al. (eds) *Handbook of Science and Technology Studies*, Thousand Oaks, CA: Sage: 343–59.

—— (1995b) 'From fax to facts: communication in the cold fusion saga', *Social Studies of Science*, 25: 403–36.

Lewontin, R. (1996) 'In the blood', *New York Review of Books*, 23 May: 31–2.

Macdonald, S. and Silverstone, R. (1992) 'Science on display: the representation of scientific controversy in museum exhibition', *Public Understanding of Science*, 1: 69–87.

Mazur, A. (1981) 'Media coverage and public opinion on scientific controversies', *Journal of Communication*, 31: 106–15

Michael, M. (2002) 'Comprehension, apprehension, prehension: heterogeneity and the public understanding of science', *Science Technology & Human Values*, 27: 357–78.

Miller, S. (1994) 'Wrinkles, ripples and fireballs: cosmology on the front page', *Public Understanding of Science*, 3: 445–53.

Moscovici, S. (1961) *La psychanalyse, son image, son public*, Paris: Puf.

Neresini, F. (2000) 'And man descended from the sheep: the public debate on cloning in the Italian press', *Public Understanding of Science*, 9: 359–82.

Nowotny, H., Scott, P. and Gibbons, M. (2001) *Re-Thinking Science. Knowledge and the Public in an Age of Uncertainty*, Cambridge: Polity Press.

Pais, A. (1982) *Subtle is the Lord...: the Science and Life of Albert Einstein*, New York: Oxford University Press.

Peters, H. P.(1995) 'The interaction of journalists and scientific experts: co-operation and conflict between two professional cultures', *Media Culture & Society*, 17: 31–48.

Phillips, D. M. (1991) 'Importance of the lay press in the transmission of medical knowledge to the scientific community', *New England Journal of Medicine*, 11 Oct: 1180–3.

Raichvarg, D. and Jacques, J. (1991) *Savants et Ignorants. Une Histoire de la Vulgarisation des Sciences*, Paris: Seuil.

Reddy, M. (1979) 'The conduit metaphor. A case of frame conflict in our language about language', in Ortony, A. (ed.) *Metaphor and Thought*, Cambridge: Cambridge University Press, 284–324.

Rowe, G. and Frewer, L. J. (2005) 'A typology of public mechanisms', *Science, Technology & Human Values*, 30: 251–90.

Shinn, T. and Whitley, R. (1985) (eds) *Expository Science. Forms and Functions of Popularization*, Dordrecht: Reidel.

Stephens, L. F. (2005) 'News narratives about nano S&T in major US and non-US newspapers', *Science Communication*, 27: 175–99.

Stilgoe, J., Wilsdon, J. and Wynne, B. (2005) *The Public Value of Science*, London: Demos.

Sturgis, P. J. and Allum, N. C. (2004) 'Science in society: re-evaluating the deficit model of public attitudes', *Public Understanding of Science*, 13: 55–74.

Trench, B. (2006) 'Science communication and citizen science: how dead is the deficit model?', paper presented at PCST9 Conference, Seoul, 17–19 May 2006.

Turney, J. (1998) *Frankenstein's Footsteps. Science, Genetics and Popular Culture*, New Haven, CT: Yale University Press.

Whitley, R. (1985) 'Knowledge producers and knowledge acquirers', in Shinn, T. and Whitley, R. (eds), *Expository Science. Forms and Functions of Popularization*, Dordrecht: Reidel, 3–28.

Wynne, B. (1989) 'Sheepfarming after Chernobyl: a case study in communicating scientific information', *Environment Magazine*, 31: 10–39.

—— (1995) 'Public understanding of science', in Jasanoff et al. (eds) *Handbook of Science and Technology Studies*, Thousand Oaks, CA: Sage: 361–89.

Ziman, J. (2000) *Real Science. What It Is, and What It Means*, Cambridge: Cambridge University Press.

Zola, E. (1871) *La Fortune des Rougon* (edn 1981, Flammarion, Paris).

6

Health campaign research

Robert A. Logan

Introduction

This chapter is a literature review of the conceptual landscape upon which health campaign research is based, and discusses some recent ideas to advance scholarship in the field. It also briefly describes some common research variables. Some of the diseases, conditions and public health challenges that health communication campaign researchers have addressed are mentioned. The evolution of the conceptual frameworks underlying health communication research, and some widely used theoretical models, are introduced with accompanying challenges and opportunities for future research.

The chapter focuses on the development of research about health behaviour decisions when non-fiction mass media are used and where campaigns are supplemented by interpersonal support systems. Most of the research cited here is from the USA, where health campaign research has been active for more than 30 years. The chapter's focal point is the conceptual underpinnings of research as there have been recent overviews describing and critiquing the primary findings from health communication campaigns (Atkin 2001; Rice and Katz 2001; Hornik 2002a; Thompson et al. 2003; Dutta-Bergman 2005; Murero and Rice 2006; Noar 2006).

Intent of health campaigns

Health communication campaigns are either informative or persuasive in nature (Atkin 2001; Piotrow and Kincaid 2001). They minimally seek to change a target audience's awareness or basic knowledge about a disease or condition. For example, a prototypical health communication campaign might seek to inform medically underserved women about (a) the local availability of mammograms; and (b) the importance of screening to promote the prevention, early detection and successful treatment of breast cancer. Some campaigns also try to change a target audience's attitudes or ability to cope with a disease or condition, such as boosting the motivation of low-income Latino women to obtain a periodic mammogram.

More ambitious health communication campaigns seek to change a target audience's behavioural inclinations to take a specific action, such as by fostering specific steps to obtain a mammogram from a certified physician or community clinic. The most ambitious campaigns seek therapeutically to change specific clinical outcomes for a targeted audience. The outcomes of the latter type of study, hypothetically, might demonstrate that after controlling for possible confounding variables, the mean length for treatment declined significantly among the intervention compared with a treatment group condition (Hornik 2002a).

Hence health campaign research tries to provide evidence of individual and audience effects along a sophisticated behavioural continuum. More specifically, a continuum of intent in health communication campaigns seeks to demonstrate improvements in a targeted audience's:

- awareness and knowledge – cognition about an illness, condition or public health threat;
- attitudes about health – affective dimension, or feelings and motivations about an illness, condition or public health concern;
- behavioural inclinations – a conative dimension, or intentions to obtain assistance for an illness, condition or public health concern;
- specific therapeutic health-related, behavioural actions;
- specific clinical treatment outcomes (Fishbein and Ajzen 1975; Piotrow et al. 1997; Wimmer and Dominick 2006).

To public health officials and healthcare providers, well orchestrated health campaigns are a therapeutic intervention designed to improve personal health and patient education, or a constructive strategy to inform or influence public health (Pettegrew and Logan 1987). The health communication campaigns reviewed here mostly deliver non-fiction messages through mass media delivery systems (radio, television, newspapers, magazines, brochures, direct mail, billboards, posters, etc.) to publicise a desired health behaviour. Public health campaigns are usually non-commercial health education efforts. Health campaigns are a subset of a much larger academic field, public health communication, and marketing that includes areas such as direct-to-consumer pharmaceutical commercial advertising (Thompson et al. 2003).

While health communication campaigns can be as simple as a solo media intervention, a campaign's publicity is often accompanied by parallel patient and public educational efforts provided by healthcare institutions, such as hospitals and managed care organisations, and government agencies. These are often supplemented by interpersonal support from healthcare providers, or by trusted local sources such as schools, religious institutions and community-based organisations, that reinforce the targeted health behaviours in the mass media campaign (Flay and Burton 1988).

Research approaches and common variables

In terms of research, health communication campaigns are often a hybrid of social science and clinical intervention approaches regarding study design, implementation and evaluation. For example, the Stanford heart health research (one of the field's pioneering studies in the 1970s) featured social science-derived theories, and compared intervention treatment and control groups (Farquhar 1983; Farquhar et al. 1984; Flora 2001). The Stanford heart health research treatments included a solo mass media campaign promoting the prevention of heart disease; the same media campaign accompanied by orchestrated reinforcement from participating healthcare providers and community healthcare organisations; plus a control community where no intervention occurred (Farquhar et al. 1981, 1984; Farquhar 1983; Flora 2001).

Some of the independent variables used in health communication campaign research (which predate the Stanford study) include demographic measures such as age, income, gender and educational level. Today, these are sometimes coupled with non-demographic measures and intermediate variables such as a person's perceived health status, or a respondent's health information-seeking habits from informal sources. Informal sources include the mass media plus family and friends; formal sources include healthcare providers. Self-efficacy (a person's self-perceived ability to implement the healthcare recommendation suggested by the campaign) is another commonly used intermediate variable (Bandura 1995). Some common dependent variables reflect the continuum described above, and might include operationalisations of audience: awareness, attitudes, knowledge, behavioural intentions, behavioural changes, and measures of specific clinical outcomes.

Some standard demographic outcome variables (employment, marital status, income, race/ethnicity) and group assignments (age, gender, native language, internet use) are included within the US National Cancer Institute's Health Information National Trends Survey (HINTS, http://cancercontrol.cancer.gov/hints). Other outcome variables within HINTS include media exposure, exposure to health information, trust in information sources, internet use, plus variables regarding health status, social networks, risk perception, disease representations (perceptions of cancer prevention, early detection and treatment), cancer communication experiences, history, information-seeking behaviour, efficacy and use of information sources. Although HINTS focuses on cancer-related issues, it supplies a statistically valid and reliable survey of more than 6300 Americans. HINTS was launched in 2003 and a second survey was conducted in 2005. While it does not assess the impact of a campaign, HINTS provides baseline consumer data and is a free resource that encourages secondary analysis. Its utility was illustrated recently in a series of articles in the *Journal of Health Communication* (Hesse et al. 2006; Viswanath et al. 2006).

While the focus of this chapter is on influencing consumer behavioural decisions, Maibach and Holtgrave (1995) note that other types of health communication campaign generate different outcome variables than those outlined above. For example, a campaign might assess the impact of media health advocacy with a selected audience of consumers or journalists. Other independent or intermediate variables might include the effectiveness of different types of campaign designs, such as assessing the impact of fear appeals within targeted messages. Other independent or intermediate variables might assess the impact of various types of risk communication strategies within targeted messages. In lieu of the macro-conceptual issues discussed in this chapter, Noar (2006) believes that the field's narrower challenge is to increase research about effective campaign design, or how different types of health message and communication strategy affect the continuum of intended outcomes mentioned above.

In contrast to research that relies on non-fiction mass media, some campaigns depict health information within a fictional entertainment media setting, such as placing non-fiction health messages within a televised, mass audience situation comedy or film (DeJong and Winsten 1990). The extension of health campaign research to all types of mass media delivery systems, and the uninhibited use of persuasion-inspired mass communication strategies drawn from all forms of advertising, strategic communication, public health and commercial marketing, is sometimes referred to as social marketing (Maibach and Holtgrave 1995; Langill 2004). The use of more persuasive mass communication strategies presumes that, regardless of a campaign's goals, educating and informing the public via traditional methods such as publicising medical research or information by cooperating with the news media, or encouraging health education via public schools and adult and provider–patient interaction, is not always optimal. While a review of the social marketing literature is beyond the scope of this chapter, Randolph and Viswanath (2004) note that innovative health communication approaches are needed because of increased competition for time and space on traditional mass media, and the emergence of interpersonal communication networks on the Web (blogs, listservs, email, text messaging).

In addition to focusing on quantitatively derived variables, health campaign research also has a tradition of using qualitative approaches (Dervin and Frenette 2001). For example, Dervin's sense-making theoretical model is an alternative, qualitative approach to conceiving and assessing how an individual responds to health communication campaigns. Some health communication campaigns that used qualitative research approaches are reviewed by Piotrow and Kincaid (2001), and Dutta-Bergman (2005). The use of the qualitatively based diffusion-of-innovations tradition in the design and evaluation of health communication campaigns is discussed by Backer et al. (1992), and Haider and Kreps (2004).

Whether or not a health communication campaign seeks to inform or persuade, or if its methodological approach is grounded in qualitative, quantitative, or mixed methods, the field is multidisciplinary and often borrows research tools, communication strategies, theoretical frameworks, research methods and outcome variables from diverse disciplines. These include cognitive psychology, social psychology, public opinion, public health, mass communication, communication studies, strategic communication, secondary and higher education, health education, advertising, social marketing and, more recently, consumer health informatics.

What do health campaigns address?

In a meta-analysis of 48 social science-based health communication campaigns that roughly equate to the non-fiction-based, quantitative research designs noted above, Snyder (2001) found that most campaigns either attempt to persuade people to stop an existing health behaviour, or promote the commencement of a new therapeutic behaviour.

Health communication campaigns designed to change counterproductive behaviours may focus on adult and youth smoking, illegal drug use, alcohol abuse, binge drinking, infant sleeping positions, sex with risky partners, and alcohol sales to minors (Snyder 2001).

Some health communication campaigns intend to initiate or promote a new therapeutic behaviour. These include promotion of condom use, diet, exercise, seat belt and bike helmet use, mammography, dental visits, health status screenings, hypertension control and pap smears. Other types of campaign to initiate a new therapeutic behaviour include crime prevention, reducing heart disease, and promoting awareness of stroke's risk factors.

As mentioned, health campaign research can be classified by initiatives that address a disease/condition or a public health concern. Among the diseases and conditions frequently associated with health communication campaigns aimed to prevent chronic disease are AIDS, other sexually transmitted diseases, heart disease, stroke, breast cancer, prostate cancer, skin cancer (melanoma), cervical cancer, hypertension, diabetes, oral health and sudden infant death syndrome. Among health campaigns that focus more broadly on public health or health-promotion efforts are promoting smoking cessation, reducing alcohol consumption, seat-belt use, eating more fruits and vegetables, and safe sex. Health campaign research focuses on a variety of age groups, including children, adolescents and adults. Some campaigns are targeted at medically underserved audiences, demographic segments where the probability of risky health behaviours is higher and access to medical services is lower than in the general population (Piotrow and Kincaid 2001; Dutta-Bergman 2005, 2006).

While most of the health communication campaign research that Snyder (2001) reviewed occurred in North America and western Europe, there is increasing activity in the rest of the world, including in many developing nations (Piotrow et al. 1997; Piotrow and Kincaid 2001; Hornik 2002b; Dutta-Bergman 2005).

Health campaigns: a short conceptual history

Media-based public health communication campaigns have been common within the USA for almost three centuries (Paisley 2001). As early as 1721, the Reverend Cotton Mather spearheaded a community-based effort to promote inoculation and prevent the spread of smallpox in Boston. In the 1840s, Dorothea Dix used books, newspapers, and printed materials to crusade against the poor treatment of mental illness in New England. By the 1870s, well organised national campaigns to prevent alcoholic beverage sales used a combination of media, such as newspapers and publications, community resources (e.g. community meetings and church interventions) and the interpersonal influences of clergy, medical professionals, public officials and others to boost awareness of alcohol's health consequences. The latter was intended to influence public opinion, advocate cultural change, and alter public policy in a manner akin to the contemporary public health advocacy strategies advanced by Wallack (1993); Wallack and Dorfman (2001). While reporting about the impact of these campaigns was anecdotal, it is important to note that campaigns to improve individual and public health inherit a long tradition of non-commercial, media-based persuasion.

The beginnings of modern campaigns and the use of mass media as an integral part of orchestrated, public educational health interventions were probably fostered by the perceived success of journalists and authors to influence public opinion around the turn of the 20th century. By the early 20th century, the new muckraking tradition in American journalism first suggested that journalists (and the mass media) had the potential to galvanise public awareness, raise social consciousness and generate political pressure to foster reforms, which included improvements in public health. Although the muckrakers wrote about diverse socioeconomic issues, poor public health conditions evidenced by childhood employment, poor food safety and a neglect of occupational and urban sanitation were frequent topics. The popularity of muckraking is widely interpreted as generating pressure on American politicians to legislate a variety of public health reforms during the first third of the 20th century (Emery and Emery 1978; Weinberg and Weinberg 2001).

Tobey (1971) explains that the unprecedented interest in popularising science and health among US scientists and physicians in the first 30 years of the 20th century reflected a new acknowledgement of the capacity of the mass media (especially newspapers) to generate public attention and enthusiasm. Initial popularisation efforts by leading US scientists to work with journalists to increase the public's exposure to scientific and biomedical research reflected a mix of public-spirited and parochial motives (Tobey 1971). Scientists who worked to generate press coverage were parochially motivated to improve science's public goodwill, which was intended to boost scientific lobbying efforts and increase government funds for research (Tobey 1971). The more public-spirited motives included a sincere interest in helping people to live healthier lives through making it easier for them to learn about biomedical and scientific achievements.

Hence, from the start of popularisation, its advocates had a sincere interest in promoting social and individual learning coupled with an interest in persuasion and social influence. Taking a more critical perspective, the effort may also have reflected an attempt to boost a cultural hierarchy where approaches to addressing policy initiatives were grounded primarily in scientific knowledge (Tobey 1971).

Regardless of their intent, the efforts by popularisation's advocates reflected an underlying conceptual framework that posited a triangular positive association between (a) the persuasive power of the news media; (b) message transmission (e.g. placement of science and biomedical news and publicity in popular media); and (c) audience exposure (increasing the odds that the general public would see science and medical information by placing it within the popular media) (Logan 2001). One of the odd legacies of the development of health campaign research is that this aspect of popularisation's conceptual framework was partially adopted, and remains influential within the field.

As recently as 2001, 13 output persuasion steps in health communication campaigns (McGuire 2001) were identified as:

- tuning in (exposure to the message);
- attending to the communication;
- liking it; maintaining interest in it;
- comprehending its contents;

- generating related cognitions;
- acquiring relevant skills;
- agreeing with the communication;
- storing this new position in memory;
- retrieval of this new position from memory when relevant;
- decision to act on the position;
- acting on it;
- post-action cognitive behaviour on the position;
- proselytising others to behave similarly.

But in parallel to imagining persuasion as a simple process of transmission and exposure via popular media, health communication campaign researchers in the 1970s adopted a more sophisticated conceptual framework about mass media's processes and effects. Unlike popularisation, health communication researchers perceived the transfer of public information to be an inefficient process, with multiple barriers that potentially explained how and why an audience's responses to health messages in the media often differed from the intent of the original source (or communicator) (Pettegrew and Logan 1987; McGuire 2001).

This more sophisticated conceptual framework noted how the perception of mass media messages about health were affected by five separate, interactive dimensions – each of which contained potential barriers that could undermine the intent of a communicator, or potentially explain why health campaigns might not result in intended increases in a targeted audience's knowledge, attitudes, cognitions and eventual behaviours. A conceptual framework emerged, which noted that potential health communication barriers or constraints could be a by-product of the health messages' (a) source, (b) content, (c) media-delivery channel, (d) receiver's (individual) post-exposure, and (e) destination.

McGuire (2001) explains that some of the source-based barriers to a desired audience response regarding health messages include the number of potentially different sources to which an individual is exposed, the unanimity or consistency of healthcare messages, the perceived demographic similarity between the source and the receiver, and the source's perceived appeal and credibility.

Some of the message-based barriers include the message's perceived appeal, the inclusion or omission of pertinent facts or familiar story frames, the arrangement and organisation of text and graphics, a text's readability for its intended audience, the inclusion of appropriate images or diagrams within a text-based story, and the cultural appropriateness of the content (ibid.).

Some of the channel-based barriers include the perceived appropriate use of print, radio, video or interactive channels to convey content (ibid.).

Some of the receiver-based barriers to a desired audience response regarding health messages include differences in perception about and interest in health as a by-product of gender, age, education, ethnicity, socioeconomic status, literacy, health literacy (ability to understand health information regardless of general education), personality, lifestyle and values. Other barriers included a person's access to mass media and the healthcare-delivery system (e.g. a person's health insurance status and ability to afford care as well as actual distance from clinical care facilities) (ibid.; IOM 2004).

Some of the common destination-based barriers include timeliness of targeted audience's receipt of information, the message's fit with audience needs, whether health messages are prevention- or cessation-oriented, and whether messages urge health behaviours that have an immediate or a delayed reward (McGuire 2001).

While these five interactive dimensions and barriers seemed comprehensive, the conceptual framework underlying health communication campaigns expanded significantly as new sociocultural dimensions regarding audience receptivity to health campaign messages were identified. Three socioculturally oriented frameworks widely used in health campaign research include:

- social influences (the influence of peer pressure and commercial advertising on health behaviours) (Evans and Raines 1982, 1990; National Cancer Institute 1991; Flynn 1992);
- cognitive behavioural (the degree to which a person's problem-solving, decision-making and self-control skills influence health behaviours) (Kendall and Holon 1979; National Cancer Institute 1991; Flynn 1992);
- life skills (the degree to which a commitment to a specific health behaviour sometimes requires broader skills and individualised training to foster a healthier lifestyle) (Botvin et al. 1980; Botvin and Eng 1982; National Cancer Institute 1991).

The campaigns that use a conceptual framework based on social influences emphasise 'social inoculation'. To overcome perceived barriers to adopting therapeutic health behaviours, people are trained about the subtle influences of advertising and peer pressures. For example, to offset peer pressures to smoke, smoking resistance behaviours are modelled, role-played, and even turned into theatre productions to encourage rehearsal and reinforcement of smoking cessation behaviours (Evans and Raines 1990; National Cancer Institute 1991; Flynn 1992).

The campaigns that use a conceptual framework based on a cognitive behavioural approach emphasise improving a person's capacity to respond to health knowledge and make reasoned decisions. For example, to encourage smoking cessation, campaign initiatives train people in how to manage smoking impulses, improve self-efficacy and reward themselves for making appropriate health decisions (Kendall and Holon 1979; National Cancer Institute 1991; Flynn 1992).

The campaigns that use a conceptual framework based on life skills emphasise how a person can live a healthier life. For example, to encourage smoking cessation, a life-skills approach provides broader training about the value of diet, exercise, clinical self-examination, alcohol moderation and relaxation (Botvin et al. 1980; Botvin and Eng 1982; National Cancer Institute 1991).

In a recent critique of the field, Dutta-Bergman (2005) acknowledges that the accumulation of conceptual frameworks provides a more comprehensive explanation of barriers to audience receptivity, and provides some concrete strategies to offset audience resistance to health campaigns.

However, he notes that three of the currently most widely used theoretical models in health campaign research are not nearly as holistic or comprehensive as the conceptual frameworks outlined above (Dutta-Bergman 2005). He describes three theoretical models currently widely used in health communication campaigns: the theory of reasoned action, the health belief model and the extended parallel process model (ibid.).

Briefly, the *theory of reasoned action* emphasises that the salience of adopting a health behaviour depends on the interaction of people's attitudes, behavioural intentions and social networks, coupled with their perceptions regarding the positive, equivocal or negative outcomes of the proposed behaviour. The emphasis is on volitional adaptive behaviour, which occurs after individual reflection about the anticipated impact and outcomes of a proposed change within each person's immediate environment (ibid.).

The *health belief model* proposes that decisions about health are a function of how an individual assesses the severity, susceptibility, benefits, barriers, cues to action and self-efficacy of the proposed behaviour. In this model, there is an emphasis on how a person self-evaluates the health risks at issue and assesses the self-perceived outcomes of a prospective behavioural change (Janz and Becker 1984; Dutta-Bergman 2005).

The *extended parallel process model* argues that a series of rationalising steps occur after a person is exposed to a message about health, which influence a person's responses. This model emphasises that a person responds differently if he or she perceives that a health message is personally threatening within a scale from low to high. The lower the threat, the more a person ignores the message. If a health message is seen as moderate to highly threatening, a person considers the perceived efficacy of the recommended action (Witte 1992; Dutta-Bergman 2005).

Cappella (2006) adds that there are four theoretical approaches to researching health message effects: behaviour-change theories; individual information-processing models; message effects research; and systemic factors. Behaviour-change theories include the integrated models of behaviour change noted above. Information processing includes understanding how individual health information exposure is activated and how information is processed. Message effects include the impact of emotional appeals, tailoring, narratives and framing. Systemic factors include the sociocultural dimensions mentioned above (Cappella 2006).

Dutta-Bergman (2005) acknowledges that in health campaigns, as in all mass communication and social science research, theoretical models and approaches are more limited and applied than the broader conceptual framework on which they are based. But he emphasises that one of the core challenges in health campaign research is the development of a synthesis that helps reconcile the current gap between comprehensive conceptual frameworks and narrower theoretical models and approaches (Dutta-Bergman 2005).

Dutta-Bergman joins a list of scholars who have argued that an expanded conceptual framework infused with more sociocultural sensitivity is needed to overcome the field's most pressing challenges (Rakow 1989; Salmon 1989). Their research suggests that health campaigns have demonstrated only limited success in influencing target audience health behaviours (Rakow 1989; Salmon 1989; Snyder et al. 2004). The persuasive success of health campaigns has not advanced significantly since the 1970s Stanford heart-health studies, which reported modest, short-term improvements in targeted audience cognitions, attitudes and some inclinations to take desired behavioural actions (Rakow 1989; Salmon 1989; Dutta-Bergmann 2005). Finally, it has been suggested that health campaigns do not reach the most at-risk audiences, and campaigns may accidentally contribute to existing knowledge gaps between marginalised audiences and populations of higher socioeconomic status (Viswanath and Finnegan 1995).

Yet some international research suggests that campaigns that integrate a broader sociocultural conceptual framework seem to generate more sustainable interest, awareness, and attitudinal change among target audiences, and are better accepted by underserved audiences (Dutta-Bergmann 2005).

> The new direction evident in much of the growing research on community-based campaigns puts importance on acknowledging marginalized people's capability to determine their own choices, model their own behaviours, and develop epistemologies based on self-understanding.
>
> (ibid. 119)

It remains to be seen whether newer models based on sociocultural conceptual frameworks will reinvigorate the field and help overcome some of the current barriers in health campaign research to influence health behaviours. Yet there is no question that the expansion of theoretical models, approaches and conceptual frameworks is an important consideration for health campaign researchers.

The internet and health campaign research

In addition to the expansion of models and conceptual frameworks into sociocultural dimensions, another promising area in health campaigns seeks to revitalise the assistance that health communication gives to consumers, patients, care-givers, healthcare providers and medical institutions through the internet. Use of the internet to provide health information and interactive communication is often called e-health or consumer health informatics (Kreps 2001; Rice and Katz 2001; Murero and Rice 2006). While this area seems more applied than conceptual, use of the internet as a tool to advance public health education, patient interaction, consumer satisfaction and patient information-seeking represents a significant development in health campaign research.

Historically, most health campaigns have been targeted for mass audiences rather than tailored for individuals. Most health campaigns are periodic, focused efforts to impact audience health behaviours, and communication initiatives are funded for relatively limited periods. The internet opens up the possibility that health campaigns and health communication will become a routine part of patient care as well as institutional efforts to promote preventive medicine (Gerber and Eiser 2001; Kreps 2001; Gustafson et al. 2002; Forkner-Dunn 2003; Neuhauser and Kreps 2003; Wofford et al. 2005).

In terms of its inherent characteristics as a mass medium, the internet fosters individual participation, enables customisation of information by users, and provides both instant consumer access and a search capacity to manage large stores of health information (Rothert et al. 2006). The internet uniquely blends all legacy media (audio, video, photography, print, display and multimedia) to make a learning or web-surfing experience potentially more compelling for the user. But in contrast to legacy media, the internet also gives a potential 'voice' to the user by enabling direct communication to mass and specific audiences with significantly lower infrastructure-related start-up costs. This eases the process for individuals to publish, broadcast or communicate to large audiences or communities of common interest, such as persons with a similar medical disease or condition (Kreps 2001; Neuhauser and Kreps 2003). Hence the internet enables unprecedented opportunities for interaction, participation and individual learning.

After a lethargic start among healthcare providers, e-health is being adopted more widely because of the public's growing affinity for internet use alongside a series of other important developments (Forkner-Dunn 2003; Wofford et al. 2005). For consumers, in addition to increasing use of the internet for disease/condition support groups, there is growing personal access to the internet; evidence that people routinely use the internet to search for health information; and the unprecedented availability of large databases of health information designed for the general public, such as WebMD.com or MedlinePlus (Barnett and Hwang 2006; Pew Internet and American Life Project 2006). For the first time in history, most current medical literature is freely available to the public through www.pubmed.gov.

The US healthcare-delivery system is slowly moving from disease treatment to preventive mode. This increases the responsibility for providers to promote patient self-management and to provide more accessible health information to the public (US HHS 2000). While other driving forces within the healthcare-delivery system are too numerous to mention here, one catalyst is the increasing need for providers to have access to electronic patient records in order to curb medical errors within hospitals and clinics (IOM 2006). As both patients and providers may need to have access to electronic patient records, this development requires the development of sophisticated, interactive internet portals with e-health applications at hospitals, clinics and other healthcare organisations.

The urgency among providers to respond to these new demands by patients and public on the US healthcare-delivery system is seen in recent efforts to conduct research about 'health literacy'. The health literacy movement, which was initiated by the US Institute of Medicine (IOM 2004) and the US-based Joint Commission (2007), attempts to help healthcare providers and medical institutions to respond better to the new demands of a world where a dialogue with patients, caregivers and the public is a normal part of sound clinical practice and medical institutional responsibilities.

As a result, the development of e-health seems to be forcing healthcare organisations and providers to commit to a more multidimensional communication environment that includes provider–patient, patient–patient, institution–provider and institution–patient interactions, as well as unprecedented consumer access to health information.

In turn, health campaigns may no longer be part of external support to patient care or an adjunct to patient education, broader health education, and medical organisations' interactions with their clients. Health communication and campaigns may become a more integral component of patient care (e.g. providing tailored health information) and health services delivery, such as encouraging interactions among patients, caregivers and providers (Mittman and Cain 2001; Gustafson et al. 2002; Nelson and Ball 2004; Rimer and Kreuter 2006; Rothert et al. 2006).

Not surprisingly, the sudden acceptance of e-health, the use of the internet and patient/consumer/caregiver communication has resulted in an array of informatics applications across the healthcare-delivery system that are addressed by Rice and Katz (2001); Nelson and Ball (2004); Murero and Rice (2006).

The expansion of health campaigns via e-health as a routine part of clinical care may also result in increased demands to demonstrate success, including their capacity to improve patient clinical outcomes and ameliorate some of the health and information disparities that occur among medically underserved audiences.

The convergence of e-health, consumer health informatics and health communication campaign applications and research seems to be part of an emerging trend within the healthcare-delivery system to curb high costs of disease treatment and to improve patient care. This development opens up unprecedented opportunities for researchers to demonstrate that the integration of health campaigns and information helps to build relationships among and between patients, caregivers, the public, providers and medical institutions, as well as creating new research opportunities. While it is premature to assess the contribution of e-health, there is no question that the ties that bind healthcare and health communication seem to be shifting, and prospects for health campaign research may flourish.

Concluding remarks

Health campaign research offers insights into how people become engaged in learning about health and medicine, how science affects a person's attitudes and behaviours, and how people interact with mass media. Flay and Burton (1988) explain that the components of a state-of-the-art health campaign include: (a) media messages that are well tailored to the characteristics, predispositions and attitudes of the targeted audience; (b) an emphasis on interpersonal reinforcement about the goals of the campaign among targeted audience members (by trusted sources such as healthcare professionals or respected healthcare organisations); and (c) encouraging a wider dialogue about the impact of the public policy, humanitarian, ethical and sociocultural dimensions of science and health. These suggested strategies also, serendipitously, provide foundations for the public's understanding of science via the mass media (Logan 2001). As a result, health campaign research has become a test bed to explore how the public reacts to a key area of science's mass communication. Although the link between public understanding of science and health campaign research literature is under-appreciated, the two fields share interests in understanding how people converge on scientific information. Accordingly, health campaign research beckons as a venue to advance theoretical insights and applied strategies in science communication. Health campaigns assist the science communication field by providing potential sources of funding through governmental and private healthcare agencies. Health campaigns also represent a promising venue for future research.

The bibliographic contributions of Marcia Zorn, MA, US National Library of Medicine, are gratefully acknowledged.

Selected further reading

Atkin, C. K. (2001) 'Theory and principles of media health campaigns', in Rice, R. E. and Atkin, C. K. (eds) *Public Communication Campaigns*, 3rd edn, Thousand Oaks, CA: Sage, 49–68.

Dutta-Bergman, M. (2005) 'Theory and practice in health communication campaigns: a critical interrogation', *Health Communication*, 18: 103–22.

Hornik, R. C. (ed.) (2002a) *Public Health Communication: Evidence for Behavior Change*, Mahwah, NJ: Lawrence Erlbaum.

Noar, S. M. (2006) 'A 10-year retrospective of research in health mass media campaigns: where do we go from here?', *Journal of Health Communication*, 11: 21–42.

Randolph, W. and Viswanath, K. (2004) 'Lessons learned from public health mass media campaigns: marketing health in a crowded media world', *Annual Review of Public Health*, 25: 419–37.

Other references

Atkin, C. K. (2001) 'Theory and principles of media health campaigns', in Rice, R. E. and Atkin, C. K. (eds) *Public Communication Campaigns*, 3rd edn, Thousand Oaks, CA: Sage, 49–68.

Backer, T. E., Rogers, E. M. and Sopory, P. (1992) *Designing Health Communications Campaigns: What Works?*, Newbury Park, CA: Sage.

Bandura, A. (1995) 'Exercise of personal and collective efficacy', in Bandura, A. (ed.) *Self-efficacy in Changing Societies*, New York: Cambridge University Press, 1–45.

Barnett, G. A. and Hwang, J. M. (2006) 'The use of internet for health information and social support: a content analysis of online breast cancer discussion groups', in Murero, M. and Rice, R. E. (eds) *The Internet and Health Care: Theory, Research and Practice*, Mahwah, NJ: Lawrence Erlbaum.

Botvin, G. J. and Eng, A. (1982) 'The efficacy of a multicomponent approach to the prevention of cigarette smoking', *Preventive Medicine*, 11: 199–211.

Botvin, G. J., Eng, A. and Williams, C. L. (1980) 'Preventing the onset of cigarette smoking through life skills training', *Preventive Medicine*, 9: 135–43.

Cappella, J. N. (2006) 'Integrating message effects and behavior change theories: organizing comments and unanswered questions', *Journal of Communication*, 56: S265-79.

DeJong, W. and Winsten, J. A. (1990) 'The use of mass media in substance abuse prevention', *Health Affairs*, 9: 30–46.

Dervin, B. and Frenette, M. (2001) 'Sense-making methodology: communicating communicatively with campaign audiences', in Rice, R. E. and Atkin, C. K. (eds) *Public Communication Campaigns*, 3rd edn, Thousand Oaks, CA: Sage, 69–87.

Dutta-Bergman, M. (2005) 'Theory and practice in health communication campaigns: a critical interrogation', *Health Communication*, 18: 103–22.

—— (2006) 'Media use theory and internet use for health care', in Murero, M. and Rice, R. E. (eds) *The Internet and Health Care: Theory, Research and Practice*, Mahwah, NJ: Lawrence Erlbaum, 83–105.

Emery, E. and Emery, M. (1978) *The Press in America: An Interpretive History of the Mass Media*, Englewood Cliffs, NJ: Prentice Hall.

Evans, R. I. and Raines, B. E. (1982) 'Control and prevention of smoking in adolescents: a psychosocial perspective', in Coats, T. J., Peterson, A. C. and Perry, C. (eds), *Promoting Adolescent Health: A Dialog on Research and Practice*, New York: Academic Press.

—— (1990) 'Applying a social psychological model across health promotion interventions: cigarettes to smokeless tobacco', in Edwards, J., Tindale, R. S., Heath, L. and Posavac, E. J. (eds) *Social Influence Process and Prevention: Social Psychological Applications of Social Issues*, New York: Plenum Press.

Farquhar, J. W. (1983) 'Changes in American lifestyle and health', in Hamner II, J. and Jacobs, B. (eds) *Marketing and Managing Health Care: Health Promotion and Disease Prevention*, Memphis, TN: University of Tennessee Center for the Health Sciences.

Farquhar, J. W., Fortman, S., Wood, P. and Haskell, W. (1981) 'Communication studies of cardiovascular disease prevention', in Kaplan, N. and Stander, J. (eds) *Prevention of Coronary Heart Disease: Practical Management of Risk Factors*, Philadelphia, PA: Saunders.

Farquhar, J. W., Maccoby, N. and Solomon, D. (1984) 'Community applications of behavioral medicine', in Gentry, W. D. (ed.) *Handbook of Behavioral Medicine*, New York: Guilford Press.

Fishbein, M. and Ajzen, I. (1975) *Belief, Attitude, Intention and Behavior: An Introduction in Theory and Research*, Reading, MA: Addison-Wesley.

Flay, B. R. and Burton, D. (1988) 'Effective mass communication campaigns for public health', paper presented to the Conference for Mass Communications and Public Health, Rancho Mirage, CA, USA.

Flora, J. A. (2001) 'The Stanford community studies: campaigns to reduce cardiovascular disease', in Rice, R. E. and Atkin, C. K. (eds) *Public Communication Campaigns*, 3rd edn, Thousand Oaks, CA: Sage, 192–213.

Flynn, B. S. (1992) 'Prevention of cigarette smoking through mass media intervention and school programs', *American Journal of Public Health*, 82: 827–34.

Forkner-Dunn, J. (2003) 'Internet-based patient self-care: the next generation of health care delivery', *Journal of Medical Internet Research*, 5: e8.

Gerber, B. S. and Eiser, A. R. (2001) 'The patient–physician relationship in the internet age: future prospects and the research agenda', *Journal of Medical Internet Research*, 3: e15.

Gustafson, D. H., Hawkins, R. P., Boberg, E. W., McTavish, F. M., Owens, B., Wise, M., Berhe, H. and Pingree, S. (2002), 'CHESS: 10 years of research and development in consumer health informatics for board populations including the underserved', *International Journal of Medical Informatics*, 65: 169–77.

Haider, M. and Kreps, G. L. (2004), 'Forty years of diffusion of innovations: utility and value in public health', *Journal of Health Communication*, 9:S3–S11.

Hesse, B. W., Moser, R. P., Rutten, L. J. and Kreps, G. L. (2006), 'The health information national trends survey: research from the baseline', *Journal of Health Communication*, 11: svii–sxvi.

Hornik, R. C. (ed.) (2002a) *Public Health Communication: Evidence for Behavior Change*, Mahwah, NJ: Lawrence Erlbaum.

Hornik, R. C. (2002b) 'Communication in support of child survival: evidence and explanations from eight countries', in Hornik, R. C. (ed.) *Public Health Communication: Evidence for Behavior Change*, Mahwah, NJ: Lawrence Erlbaum, 219–48.

IOM (2004) *Health Literacy: A Prescription to End Confusion*, Washington, DC: Institute of Medicine of the National Academies/National Academies Press.

IOM (2006) *Preventing Medication Errors: Quality Chasm Series*, Washington, DC: Institute of Medicine of the National Academies/National Academies Press.

Janz, N. K. and Becker, M. H. (1984) 'The health belief model: a decade later', *Health Education Quarterly*, 11: 1–47.

Joint Commission (2007) *What Did the Doctor Say? Improving Health Literacy to Project Patient Safety*, Oakbrook Terrace, IL: The Joint Commission. www.jointcommission.org/NR/rdonlyres/D5248B2E-E7E6-4121-8874-99C7B4888301/0/improving_health_literacy.pdf

Kendall, P. C. and Holon, S. D. (1979) *Cognitive–Behavioral Intentions: Theory, Research and Practice*, New York: Academic Press.

Kreps, G. L. (2001) 'The importance of consumer health informatics', Keynote Speech to Symposium on Critical Issues in Consumer Informatics, Bethesda, MD: American Medical Informatics Association.

Langill, D. (ed.) (2004) *Selling Health Lifestyles: Using Social Marketing to Promote Change and Prevent Disease.* Issue brief. Washington, DC: Grantmakers in Health.

Logan, R. A. (2001) 'Science mass communication: its conceptual history', *Science Communication*, 23: 135–63.

Maibach, E. and Holtgrave, D. R. (1995) 'Advances in public health campaigns', *Annual Review of Public Health*, 16: 219–38.

McGuire, W. J. (2001) 'Input and output variables currently promising for construction persuasive communications', in Rice, R. E. and Atkin, C. K. (eds) *Public Communication Campaigns*, 3rd edn, Thousand Oaks, CA: Sage, 22–48.

Mittman, R. and Cain, M. (2001) 'The future of the internet in health care: a five-year forecast', in Rice, R. E. and Katz, J. E. (eds) *The Internet and Health Communication: Experiences and Expectations*, Thousand Oaks, CA: Sage, 47–74.

Murero, M. and Rice, R. E. (eds) (2006) *The Internet and Health Care: Theory, Research and Practice*, Mahwah, NJ: Lawrence Erlbaum.

National Cancer Institute (1991) *Strategies to Control Tobacco Use in the United States: A Blueprint for Public Health Action in the 1990s*, Washington, DC: National Cancer Institute.

Nelson, R. and Ball, M. J. (2004) *Consumer Informatics: Applications and Strategies in Cyber Health Care*, New York: Springer.

Neuhauser, L. and Kreps, G. (2003) 'Rethinking communication in the e-health era', *Journal of Health Psychology*, 8: 7–22.

Noar, S. M. (2006) 'A 10-year retrospective of research in health mass media campaigns: where do we go from here?', *Journal of Health Communication*, 11: 21–42.

Paisley, W. J. (2001) 'Public communication campaigns: the American experience', in Rice, R. E. and Atkin, C. K. (eds) *Public Communication Campaigns*, 3rd edn, Thousand Oaks, CA: Sage, 3–21.

Pettegrew, L. and Logan, R. A. (1987) 'Health communication: review of theory and research', in Berger, C. R. and Chaffee, S. H. (eds) *Handbook of Communication Science*, Beverly Hills, CA: Sage, 675–710.

Pew Internet and American Life Project (2006), *Reports: Health*, www.pewinternet.org/PPF/r/190/report_display.asp

Piotrow, P. T. and Kincaid, D. L. (2001) 'Strategic communication for international health programs', in Rice, R. E. and Atkin, C. K. (eds) *Public Communication Campaigns*, 3rd edn, Thousand Oaks, CA: Sage, 249–266.

Piotrow, P. T., Kincaid, D. L., Rimon II, J. G. and Rinehart, W. (1997) *Health Communication: Lessons from Family Planning and Reproductive Health*, Westport, CN: Praeger.

Rakow, L. F. (1989) 'Information and power: towards a critical theory of information campaigns', in Salmon, C. (ed.) *Information Campaigns: Balancing Social Values and Social Change*, Newbury Park, CA: Sage.

Randolph, W. and Viswanath, K. (2004) 'Lessons learned from public health mass media campaigns: marketing health in a crowded media world', *Annual Review of Public Health*, 25: 419–37.

Rice, R. E. and Katz, J. E. (eds) (2001) *The Internet and Health Communication: Experiences and Expectations*, Thousand Oaks, CA: Sage.

Rimer, B. and Kreuter, M. W. (2006) 'Advancing tailored health communication: a persuasion and message effects perspective', *Journal of Communication*, 56: S184–S201.

Rothert, K., Strecher, V. J., Doyle, L. A., Caplan, W. M., Joyce, J. S., Jimison, H. B., Karm, L. M., Mims, A. D. and Roth, M. A. (2006) 'Web-based weight management programs in an integrated health care setting; a randomized, controlled trial', *Obesity*, February 14: 266–72.

Salmon, C. (ed.) (1989) *Information Campaigns: Balancing Social Values and Social Change*, Newbury Park, CA: Sage.

Snyder, L. B. (2001) 'How effective are medicated health campaigns?' in Rice, R. E. and Atkin, C. K. (eds) *Public Communication Campaigns*, 3rd edn, Thousand Oaks, CA: Sage, 181–92.

Snyder, L. B., Hamilton, M. A., Mitchell, E. W., Kiwanuka-Tondo, J., Fleming-Milici, F. and Proctor, D. (2004) 'A meta-analysis of the effect of mediated health communication campaigns on behavior change in the United States', *Journal of Health Communication*, 9: S71–S96.

Thompson, T. L., Dorsey, A. M., Miller, K. I. and Parrot, R. (eds) (2003) *Handbook of Health Communication*, Mahwah, NJ: Lawrence Erlbaum.

Tobey, R. (1971) *The American Ideology of National Science*, Pittsburgh, PA: University of Pittsburgh Press.

US HHS (2000) *Healthy People 2010*. Conference edn in two volumes. Washington, DC: US Department of Health and Human Services/US Government Printing Office.

Viswanath, K. and Finnegan, J. R. (1995) 'The knowledge gap hypothesis: twenty-five years later', in Burlson, B. (ed.) *Communication Yearbook 19*. Thousand Oaks, CA: Sage, 187–228.

Viswanath, K., Breen, N., Meissner, H., Moser, R., Hesse, B., Steele, W. and Rakowski, W. (2006) 'Cancer knowledge and disparities in the information age', *Journal of Health Communication*, 11: S1–S17.

Wallack, L. (1993) *Media Advocacy and Public Health: Power for Prevention*, Newbury Park, CA: Sage.

Wallack, L. and Dorfman, L. (2001) 'Putting policy into health communications: the role of media advocacy', in Rice, R. E. and Atkin, C. K. (eds) *Public Communication Campaigns*, 3rd edn, Thousand Oaks, CA: Sage, 389–402.

Weinberg, A. and Weinberg, L. (2001) *The Muckrakers*, Urbana, IL: University of Illinois Press.

Wimmer, R. D. and Dominick, J. R. (2006) *Mass Communication Research: An Introduction*, 8th edn, Belmont, CA: Wadsworth.

Witte, K. (1992) 'The extended parallel process model', *Communication Monographs*, 59: 329–49.

Wofford, J. L., Smith, E. D. and Miller, D. P. (2005) 'The multimedia computer for office-based patient education: a systematic review', *Patient Education and Counseling*, 59: 148–57.

7

Genetics and genomics

The politics and ethics of metaphorical framing

Iina Hellsten and Brigitte Nerlich

Introduction

This chapter focuses on the use of metaphors as common points of reference that establish relationships between the sciences, the mass media, and their publics. In particular, we discuss public debates about genetics and genomics from the 1990s to the 2000s in order to offer insights into the politics and ethics of metaphorical framing. Our argument is that metaphors, such as 'clones are copies', play an important role in the public communication of science and technology, just as they do in science itself. As Dupré pointed out, 'it has long been argued that all science depends on metaphors. Understanding grows by the projection of a framework through which we understand one kind of thing onto some less familiar realm of phenomena' (Dupré 2007). Metaphors in science have constitutive, explanatory and communicative functions. In this chapter we focus on the communicative function of metaphors in public debates about science and technology.

For more than half a century, public debates on developments in biotechnology in particular, and life sciences in general, have been dominated by two narrative frames. On the one hand, advances in genetics and genomics have been covered in terms of sensational *breakthroughs* in the progress of a science the aim of which is often said to be to *reveal the secrets of the book of life* and to provide a *key* to curing common diseases. On the other hand, specific applications of biotechnology, such as cloning and stem cell research, have been framed as scientists *playing God*, opening *Pandora's Box*, and creating *Frankenstein's monsters*. Both narratives are grounded in a view of scientific and technological progress as a linear movement in space, a *journey* either to *map* unknown territory, or to enter the darker regions of the manipulation of life, the consequences of which may become monstrous. These seemingly opposite narratives on science's progress both frame scientific and technological progress in terms of a journey (Hellsten 2002: 1–3; 133–5; Ceccarelli 2004). Such entrenched metaphorical framings are not easy to shift and can blind our imagination to other possible ways of grasping developments in science and technology.

The rhetoric of great promises of biotechnological research reached a climax in June 2000, in the public announcement that the sequencing and mapping of the human genome were nearing completion (Hellsten 2001; Nerlich et al. 2002; Nerlich and Dingwall 2003; Calsamiglia and van Dijk 2004; Ceccarelli 2004; Nerlich and Hellsten 2004; Gogorosi 2005; Bostanci 2006; Henderson and Kitzinger 2007). In the mass media, this scientific 'achievement' was glorified as the *opening of the book of life*, the revelation of the *secret text of life* or the finding of the *key* to curing diseases. The human genome itself was discussed as *a wondrous map*, a completed *book of life* or the *blueprint* for a person. These metaphors bathed the Human Genome Project (HGP) in a positive light that was intended to reflect back on the status of genetics and genomics at large, a field that had also revealed what some might regard as the darker and more worrying secrets of cloning and embryonic stem cells at the end of the 1990s. However, this metaphor of science progressing or, indeed, racing through a series of breakthroughs seems to imply a view of science as standing apart from society at large, and to strengthen the idea of science communication as one-way transfer of news of such scientific achievements to a grateful public.

The aim of this chapter is to give an overview of research on the role of metaphors in the public communication of science and technology, using genetics and genomics as the point of reference, and to discuss the role of specific types of metaphor, 'discourse metaphors' (Zinken et al. 2003, 2008), in such communication. We ask in this chapter: what role do (discourse) metaphors play in the public communication of genetics and genomics? How are these metaphors connected to the framing of scientific advances as a series of breakthroughs? Whose images and purposes do such metaphors support? And what are the political and ethical implications of such metaphorical framings?

Metaphors and science communication

Metaphor research is carried out in a wide range of fields, from poetry (Hawkes 1972) to archaeology (Mithen 1998). In general, metaphors are defined as interaction between a source and a target domain, or a mapping between these two domains (Black 1962; Hesse 1988; Lakoff and Johnson 1980; Richards 1989). In the metaphor 'life is a journey', for example, life is the target domain and journey the source domain, and the mapping from life to journey consists of a selected set of structural elements that are expected to be familiar to the users. Which elements of the source domain are used in discussing the target domain depends on the purposes of the user as well as the wider context of use (Chiappe 1998; Hellsten 2000). One specific source domain, journey, for instance, can be mapped onto various target domains, such as life, love or, indeed, science. Journey as a source domain provides a perspective on science that reduces the complexity of the object (Burke 1966) and makes it amenable to understanding.

In research concerned with science communication, metaphors have been investigated in two contexts. First, metaphors are used in communication within the sciences (Montuschi 1995; Hallyn 2000) as important for the creation of new ideas (Hesse 1966; Leatherdale 1974; Bono 1990; Schön 1993), for building novel scientific models

(Black 1962; Knorr 1981), and for transferring ideas and terms across scientific fields of research, for example, from clockworks to astronomy (Newton's understanding of the universe) and from astronomy to atomic physics (Bohr's model of the atom). Second, scientists use metaphors in communicating their work to wider audiences (Bucchi 1998; Van Dijck 1998). In the mass media, metaphors are used as part of journalistic routines for purposes of popularising, concretising and dramatising issues, that is, for making issues both newsworthy and interesting for the relevant audiences (Anton and McCourt 1995). As Bucchi (1998: 30) notes, metaphors are also used to address different audiences simultaneously. The ability of the mass media to produce and sell news, and therefore to address a diverse public, is dependent on such resonance: the news has to resonate with something familiar, with earlier, accessible ways of framing issues (Gamson and Modigliani 1989; Benford and Snow 2000) and across different topics. In this process, metaphors offer a way of understanding new issues and complex processes in terms of shared experiences. In addition, metaphors may be used to evoke powerful images and emotions, thus adding drama to the news (Väliverronen 1998).

Another research tradition has perceived any concept as a potential metaphor; thus 'metaphoricity' can be defined as function of a concept in its context of use (Bono 1990; Maasen and Weingart 2000). In their study of the role of metaphors in the dynamics of knowledge, Maasen and Weingart (2000) followed the terms 'paradigm', 'chaos' and 'struggle for existence' across various disciplines over time as opening up common ground for debates. Metaphors such as 'science is a journey' may serve as communicative tools, connecting various discourses and offering common ground for debate (Maasen and Weingart 2000). They act like boundary objects that are both flexible enough to adapt to novel situations, and robust enough to maintain identifiable structures (Star et al. 1989). Like social representations in general, they are ambiguous and therefore adaptable enough to allow several uses and interpretations (Moscovici 1984), both over time and across various topics in a society, without necessarily losing their original implications.

A more recent line of research has focused on discourse metaphors that are 'relatively stable metaphorical mappings that function as a key framing device within a particular discourse over a certain period of time' (Zinken et al. 2008). Discourse metaphors evolve over time in the discourses where they are used, and resonate over time, across topics, across actors, and between each other. This line of research focuses on the dynamics of metaphors, and argues that metaphors and their meanings tend to become conventional in use and in interaction with other related metaphors and sociocultural contexts. Discourse metaphors are emergent and flexible rather than being universal conceptual structures.

In this process, previous uses of a particular source domain, which might have been very useful in generating metaphorical mappings at certain periods, become entrenched and may limit users' ability to explore different source domains when creating new metaphors. The creation and understanding of metaphors in actual discourse is therefore constrained by existing conceptual mappings and discourse metaphors, the history of metaphor use, and social, political and cultural preferences.[1] The myths and metaphors around 'man' playing God, or the creation of 'monsters', for example, have their roots in a long cultural history (e.g. the Prometheus myth), which may give such metaphors a special appeal and stability. Further, metaphors form complex families of linked metaphors, relatively stable ecologies of metaphors (Nerlich and Dingwall 2003) that have different life cycles in different domains of use. Such metaphors can be 'chosen by speakers to achieve particular communication goals within particular contexts rather than being *predetermined* by bodily experience' (Charteris-Black 2004: 247). Whereas in the conceptual metaphor tradition the focus is on the 'embodiment' of metaphors, in the communicational metaphor tradition the focus is on sociocultural 'embeddedness'.

The following section discusses the public communication of genetics and genomics as a case to illustrate the communicative roles of discourse metaphors in the public debate about science and technology.

The Human Genome Project

On 26 June 2000, a group of scientists and politicians announced in a fanfare of publicity that the human genome was nearly mapped. Bill Clinton, then President of the USA, held a press briefing at the White House flanked by the leaders of the two competing US human genome projects, Dr Craig Venter of the private company Celera Genomics, and Dr Francis Collins of the publicly funded International Human Genome Project. They were joined via satellite link from London by Tony Blair, Prime Minister of the UK. In the press briefing, President Clinton declared:

> Nearly two centuries ago, in this room, on this floor, Thomas Jefferson and a trusted aide spread out a *magnificent map* – a map Jefferson had long prayed he would get to see in his lifetime. The aide was Meriwether Lewis and the map was the product of his courageous expedition across the American frontier, all the way to the Pacific. It was a map that defined the contours and forever expanded the frontiers of our continent and our imagination.
>
> [. . .] After all, when Galileo discovered he could use the tools of mathematics and mechanics to understand the motion of celestial bodies, he felt, in the words of one eminent researcher, 'that he had learned the language in which *God created the universe*'.
>
> Today, we are learning *the language in which God created life*. We are gaining ever more awe for the complexity, the beauty, the wonder of God's most divine and sacred gift. With this profound new knowledge, humankind is on the verge of gaining immense, new power to heal.
>
> (White House 2000 [italics added])

This declaration reverberated through mass media worldwide, and scattered the seeds of cognitively and culturally hardy metaphors that could flower in almost any language and culture (see Nerlich and Kidd 2005), such as the genome as a map and a language – not just any sort of map or language, but a map that would open up new frontiers, like the map used to conquer the Wild West; and a language that was spoken by God.

Eight months later, in February 2001, the two competing groups involved in the genome sequencing project published their results in *Nature* and *Science* (Lander et al.

2001; Venter et al. 2001). These articles confirmed one unexpected result of the genome project: the human genome may contain only about 35,000 genes[2] instead of the expected 100,000. The challenge this posed to the so-called central dogma of biology did not seem to attract much media attention. The central dogma had emerged after the discovery of the double helical structure of the DNA molecule. Francis Crick and James Watson, discoverers of the double helix (Watson and Crick 1953) and two of the most celebrated heroes of science, had explained the functioning of the genes as a linear transfer of information, where one gene codes for one protein via RNA. This idea led to what Keller (2000) has called 'the century of gene', and it had implications for claims about finding gene-specific cures to diseases once the human genome was fully mapped.

The Human Genome Project was finally declared completed on 14 April 2003, on a date that coincided with the 50th anniversary of the discovery of the double helix structure of DNA, a coincidence that reinforced images of scientific glory. However, in 2003 the HGP was no longer celebrated for revealing the language of life, but instead as laying the foundation or basis for further research. The HGP was no longer seen as achieving an ultimate goal, but rather as a stepping-stone on a much longer, but still glorious, journey of science:

> The successful completion this month of all the original goals of the HGP emboldens the launch of a new phase for genomics research, to explore the remarkable landscape of opportunity that now opens up before us.
>
> (Collins et al. 2003: 286)

How was this change in views of the HGP as the principal *key* to unlocking the secrets of life, to that of the HGP as a mere *basis* for further research, justified in public? The new vision of the future painted by Collins et al. used the old architectural metaphor of the *blueprint* (which had been in use in genetics for a long time), but gave it a new twist. The earlier metaphor of the blueprint implied that DNA, or the human genome, provided the right information to build 'a human', but now the blueprint metaphor was used to indicate that the achievements of the HGP were themselves just the blueprint for a more elaborate construction: the science of the post-genomic era. The existing metaphor of the blueprint could readily be extended to the new situation, carrying with it accumulated positive associations. The authors set out their vision of the future in terms of a three-storey house built on the foundations of the sequencing success. The three floors were labelled as genomics to biology (elucidating the structure and function of genomes), genomics to health (translating genome-based knowledge into health benefits), and genomics to society (promoting genomics to maximise benefits and minimise societal harms).

The same metaphors that had been used when the HGP was launched in the 1990s, and that were reinvigorated when the working draft of the human genome was announced in June 2000, were now extended successfully in order to justify further research in genomics, from the study of whole genomes instead of individual genes, to comparative genomics, to proteomics (Tyers and Mann 2003), the study of proteins instead of genes (as genes are in some respects 'recipes' for making proteins). In the future, these metaphors will perhaps be extended further to the study of the physiome, that is, to describe the human organism quantitatively, so that one can understand its physiology and pathophysiology (see the Physiome Project, www.physiome.org). This is the provisional end of a journey that started in the 1950s with the discovery of the double-helical structure of DNA, which became itself a positive icon for genetic research (Nelkin and Lindee 1995; Bucchi 2004).

The metaphor of genes as the *book of life* is rooted in a rather blatant metonymy, according to which the bases in DNA – adenine (A), thymine (T), cytosine (C) and guanine (G) – are labelled by their initial letters. By coincidence, this 'alphabetical' mapping, initiated by Watson and Crick (1953), occurred at a time when biology was influenced by certain advances in the science of language and in information theory. This meant that favour was given to metaphorical mappings between the study of genes and the study of language, codes and computer programmes (Jacob and Monod 1961). This influence has not diminished, but has grown with the advent of bioinformatics: DNA became a code (Condit 1999: 100–2; Kay 2000: 1–37).[3] As any randomly chosen popular primer for genetics will tell you:

> The alphabet is a code. . . . DNA is also a code . . . The DNA code uses groups of three 'letters' to make meaning. Most groups of three 'letters' code for an amino acid (some code for 'punctuation' – starts and stops). For instance, the DNA letters TGC code for an amino acid called cysteine, whereas the DNA letters TGG code for an amino acid called tryptophan. Each of these sequences of three DNA letters is called a DNA triplet, or codon. Since there are four different DNA letters (A, G, C and T), there are 4 × 4 × 4 = 64 different combinations that can be used to make a codon.
>
> (http://www.science-class.net/Lessons/Genetics/Your%20Genome.pdf)

Since the 1950s, the popular metaphors of genes as *codes*, a *blueprint* and a *map* have become central to the development of the sciences of genetics and genomics. They were not just there to communicate this science to the public in a pleasingly metaphorical way, but were indispensable to biologists in making sense of their work (Van Dijck 1998: 121). The metaphor of code was created in the 1950s, a time when the theory of information was in vogue, to understand the object of the new science of gene sequencing, and to communicate this view to the public. Condit (1999: 221) even argues that '[t]he new discoveries of molecular genetics were at first not communicable to the public. Science was mute in the public sphere until it formulated the coding metaphor'. Similarly, the metaphor of the blueprint became more popular in the public communication of genetics when the focus changed from genes to genomes as whole entities (ibid. 16). As Avise pointed out, 'many genomic metaphors have elements of truth, and each may have its time and place' (Avise 2001). More recently, the physiologist and systems biologist Denis Noble (2006: 21) has made similar claims. Different, even competing, metaphors can illuminate different aspects of the same situation, each of which may be correct even though the metaphors themselves may be incompatible. We benefit most when we recognise that. We should therefore treat the competition between metaphors differently from that between descriptions that differ empirically. Metaphors compete for insight, and for criteria such as simplicity, beauty and creativity, all of which we use over and above empirical correctness in judging scientific theories. But ultimately it is by the empirical tests that scientific theories live and die (ibid.).

By contrast, there are other metaphors, such as the *book of life*, which are less central to the science itself and more central to public communication of science. Such metaphors are useful when scientists or politicians want to 'sell' science to the public (Nelkin 1995). As Avise (2001) has pointed out, the *book of life* metaphor helped focus and sell the publicly funded HGP.

The popular metaphors used by scientists and the media to communicate about genetics and genomics illuminate diachronically and synchronically different aspects, but they all live and thrive in the same semantic and conceptual field of 'coded information' and mutually reinforce each other in science discourse and science communication discourse. This increases their allure and opens them up for ideologically based misuse, for example to argue for genetic determinism or to denigrate scientists engaged in genomics by accusing them of genetic determinism.

The BBC announced the near-completion of the human genome on its internet news page (30 May 2000):

> The blueprint of humanity, the book of life, the software for existence – whatever you call it, decoding the entire three billion letters of human DNA is a monumental achievement.

The announcement was accompanied by an image in two parts (http://news.bbc.co.uk/1/hi/in_depth/sci_tech/2000/human_genome/760893.stm): a banner entitled 'The human code crackers' projected against a band of DNA 'bar codes' and further illustrated by a stylised double helix and Leonardo da Vinci's famous image of Vetruvian man with outstretched arms in the right hand corner. Underneath was another image of the double helix imprinted on a jigsaw puzzle. Here the code metaphor takes on connotations that go well beyond the rather neutral denotation of the term 'code' in genetic textbooks. It also inserts the metaphor of the code into a whole cultural, semantic and visual field of related metaphors which, on the whole, continue to highlight various deterministic and glorifying aspects of the human genome (blueprint, book of life, software of existence, puzzle-solving, code-breaking, Leonardo da Vinci, the double helix).

However, not all metaphors are created equal. Some are more suitable than others for several interpretations and uses, and some gain more prominence than others in public debates on science. Their strength is derived, as illustrated above, from the cultural associations they evoke or exploit, and from the semantic field in which they mutually support each other. The ability of some metaphors to suggest shared images may be crucial in political and public arenas in building up links between scientific knowledge and political action (Mio and Katz 1996) on the one hand, and scientific knowledge and popular knowledge on the other (Van Dijck 1998; Nerlich et al. 2000, 2001). These are, as pointed out in the introduction to this chapter, discourse metaphors. The metaphors of the *map*, *code*, *book* and *blueprint*, which pervade genomic discourse, are all discourse metaphors that are particularly apt for science communication. Discourse metaphors have a social and cultural history, and influence social and cultural futures. They can be analysed only in a broader context, not just by pondering the possible meanings of certain terms. For instance, the way in which genes as *alphabet* functions as a metaphor depends largely on how it is interpreted as a symbol of genetics or as a particular way of dealing with it. New topics introduce new ways of speaking and arguing (new discourses), but there is also discursive continuity between different topics, both synchronically and diachronically, and this continuity manifests itself in discourse metaphors and, we would say, is essentially maintained through such metaphors.

The HGP, in particular, has exploited this metaphorical stability and continuity. The reason for this might lie in what some semanticists have called 'linguistic conservatism' and what we would like to call metaphorical conservatism. Semantic studies (Ullmann 1962: 198) have shown that language tends, in general, to be more conservative than science or culture. This means that words or, in our case, metaphors that are deeply embedded in our thinking and talking might perpetuate ideas, values and attitudes which, from the point of view of science, are outdated. 'Common sense tells us that imagination is always ahead of technology, and that our technological tools keep lagging behind. However, in the context of genomics, the opposite might be more accurate: our imaginative tools can hardly keep up with our technological innovations' (Van Dijck 1998: 198). Here we could add that these metaphors seem to carry with them an idea of science communication as a linear transfer. Metaphors have cognitive and emotive staying power, and do not shift easily in parallel with scientific and technological development once they have proven their value in science communication. A process of positive feedback or amplification between science and science communication seems to be involved here that strengthens salient metaphors.

What does this mean for a science that is itself evolving? In an article entitled 'The sociable gene', Jon Turney (2005) has reflected on the growing mismatch between the scientific understanding of the gene and its popular metaphorical representations, and on the consequences of this mismatch for science communication. He writes:

> The old metaphors for genes and genomes, whether they originate in scientific discourse or in popularization or the rhetoric of research promotion, are familiar. Readers learn about the map, the code, the Book of Life, the blueprint, the recipe, the master molecule. And they often get the message that DNA is destiny.
>
> [. . .] there does appear to be an emerging mismatch between the image of the gene in the public realm and recent scientific understanding. If it is desirable to have informed public debate about genetics and its applications, it would be helpful to align these images better.
>
> (Turney 2005: 808)

Some scientists who are also science communicators, such as Craig Venter, have tried to challenge the supremacy of old genomic metaphors, especially after the 2001 publication of the full sequence of the human genome that challenged the linear idea of genes determining proteins. This is an extract of a public debate between Collins and Venter about the discourse metaphor, 'the genome is the book of life':

> What they discovered, Collins said, is that what has become known as the 'book of life' is really three books: a history book, filled with pages of the fossil

record written in our DNA code; a parts manual pointing to the genes and proteins that create a human being; and a medical text that points to risk and disease, albeit, Collins conceded, in a 'language that we don't entirely know how to read yet.

Venter took issue with Collins' characterisation of the genome. 'I don't view this as the book of life,' he said. 'And this is not the blueprint for humanity. This is a basic set of information that codes for our proteins.' You won't find the instructions for building the heart or building the brain, Venter explained, because this information comes in several different layers.

(Gross 2001)

Others not only have contested old metaphors, but have tried to invent metaphors that emphasise the complex interaction between genes and their environment, such as the human genome as an orchestra, a genome salad (Butler 2001), or, as Avise surmised in one of the rare articles exploring new metaphors, a social collective or a miniature, cellular ecosystem – thus connecting to other common sets of metaphors in public debates, networks and webs (Avise 2001: 87). Noble (2006) has made the most concerted effort yet to shake up old metaphors of genomics and to bring a new metaphor – that of the *music of life* – to the attention of other scientists and the wider public through a popular science book. Again, this was prompted in part by the discovery of the relatively low number of genes in the human genome in 2001:

Should we be surprised that there are so few? Should we not rather be amazed at the immense range of functional possibilities that such a genome can support?

A musical analogy may be helpful here. The genome is like an immense organ with 30,000 pipes. [. . .] But the music is not itself created by the organ. The organ is not a program that writes, for example, the Bach fugues. Bach did that. And it requires an accomplished organist to make the organ perform. [. . .] If there is an organ, and some music, who is the player, and who was the composer? And is there a conductor?

(Noble 2006: 33–4)

We will have to wait and see whether the new metaphor of the *music of life*, proposed by Noble, can compete with older ones – that is, whether it can resonate with public perceptions of the human genome and the HGP which, for a long time, have been dominated by the older metaphors.

The public communication, and perhaps perception, of molecular biology and DNA research has been shaped by geographical (journey) metaphors from the 1950s onwards (Van Dijck 1998; Condit 1999), but they all have their roots in longer histories of use that predate genetics and genomics. This historical ancestry provides discursive stability. The currency of the metaphor of the *book of life* dates back to antiquity and has a long history within the Judaeo-Christian tradition, where it refers to natural, eternal and universal texts (Kay 2000: 31). This universality is God-created and thus eternal. It gained its wider and more emotive currency by being rooted in Biblical imagery surrounding the book of life as evoked in Revelation:

> And I saw the dead, great and small, standing before the throne, and books were opened. Another book was opened, which is the *book of life*. The dead were judged according to what they had done as recorded in the books.
>
> The sea gave up the dead that were in it, and death and Hades gave up the dead that were in them, and each person was judged according to what he had done.
>
> (Revelation 20: 12–13, New International Version, www.biblegateway.com [italics added])

In parallel to the *book of life*, and reinforcing it, runs the metaphor of the *book of nature* common in the history of natural sciences, where science was perceived as an effort to *read and write the book of nature*. For Galileo, to whom Clinton referred in 2000, the book of nature was written in the language of mathematics (Cohen 1994).

In summary, the change from genetics to genomics has not yet led to a rush of new metaphors; instead, the existing metaphors were extended to cover new situations (Nerlich and Hellsten 2004). These stable metaphors helped 20th century scientists and the public to cope with the various genomic revolutions – they became symbolic coping mechanisms (Wagner and Hayes 2005), rooted in well established symbolic or interpretative repertoires. They also provided continuity within discontinuity.

Although the old genomic metaphors mainly evoke hopes and background fears, their use poses an inherent danger. While metaphors may help us capture novel events in terms of familiar events, '[t]here is a concomitant risk, of course: the metaphorical constructs also limit our ability to assimilate new information and, in conventional discourse, where certain literalness prevails, they can quickly lose their suppleness and become mere props for unreflective traditionalism' (Leiss 1985: 148–9).

In the following section we focus on the ethical aspects of public communication of genetics and genomics, and the use of metaphors in this activity.

The ethics of genetics and genomics

Developments in genetics and genomics are commonly supposed to be transforming not just how scientists are likely to investigate, and doctors to treat, our future illnesses, but the very way in which society operates and the way in which we view ourselves (Brown and Webster 2004). Our idea of humanity itself is in a process of creative refiguration: this is happening both literally on the level of the 'New Science' heralded and symbolised by the HGP, but also meta-linguistically and metaphorically as concepts from the new science filter into the public domain, and as concepts from the public domain filter into the representation of the science (Nerlich and Kidd 2005). In this sense, the story of the HGP is not merely that of scientific progress discussed above, but also of the nature of the public perceptions, fears and hopes involved.

In modern, technologically advanced societies, research of whatever quality can now have a direct and swift impact on public opinion and public policy. Social scientists and discourse analysts therefore have a duty to investigate how this impact is achieved, what technological, linguistic and cultural resources are used to achieve it, and to what purpose they are used. This is especially important in the fields of genetics and genomics, as we are dealing here with the 'meaning of life' itself, and promises of treatment and cure.

Moral and ethical questions that may result from increasing competition within the sciences, between research teams and between science communicators themselves, including scientists' increased need to seek public recognition via the mass media and the internet were highlighted when claims by Woo-Suk Hwang, a Korean stem cell researcher, to have achieved major breakthroughs in cloning embryos for therapeutic purposes turned out to be false. The scandal surrounding his research highlighted not only the ethical dimension of sourcing 'material' for research, but also issues of coercion, exploitation, informed consent, peer review and the pressure from politicians and funders on scientists to keep 'breakthroughs' coming (Bogner and Menz 2006; Gottweis and Triendl 2006; Franzen et al. 2007). It also exposed some aspects of media staging and science communication, which are becoming more common in a climate of increased competition for funding where 'the public' has become an important factor in the competition and in the 'knowledge transfer' activities that are now part of any science project. Metaphors around the central image or metaphor scenario of science as a journey or race, such as *breakthrough, milestone, overcoming hurdles, moving a step closer, breaking new ground, reaching a new frontier,* were plentiful in the media coverage. These metaphors shored up the hopes of patients waiting for a cure, hopes that were sadly dashed when some aspects of Hwang's research were exposed not only as unethical, but also as fraudulent (Cyranoski 2006). Soon the metaphors of race and journey went into reverse, and the media bemoaned the scientific and ethical *setbacks, stumbling blocks* and *obstacles* that this scandal had caused and exposed. Whereas the media hype surrounding the HGP based on the age-old metaphors of the *book, map, code* and *blueprint* has, on the whole, been successful in promoting genomics, the scandal surrounding Hwang's research involving human embryonic stem cells (already a rather contentious issue) has highlighted science communication as an activity in which scientists themselves engage in various, sometimes unethical, ways to promote their research. The scandal has perhaps also dented the appeal of the science as *journey/race* metaphor, and prompted a reflection on the ethics of metaphor use in science promotion and science communication (Wolvaardt 2007). Social scientists therefore should reflect not only on the risks posed by scientific advances, but also on the risks of science communication itself, which might be heightened by various aspects of 'modernity' that impinge on the research and publication process.

Concluding remarks

Metaphors play an important role in the public communication of science and technology, contributing to public understanding and misunderstanding of sciences. In our case, the main metaphors used for communicating about genetics have been readily extended to communicating about genomics, thus building a bridge between the 'new genetics', the HGP, the new genomics and post-genomics projects. The confirmation in 2001 of the relatively low number of genes that make up the human genome rendered problematic the metaphors of genes as *codes, blueprints, maps* and *books of life*, but did not lead to these metaphors being abandoned in the public communication of genetics and genomics. In this sense, the most popular metaphors have fallen behind the developments in science as well as the theorising about science communication (Turney 2005). The gap between developments in science and public representations of science may even be widening. However, a clean break in metaphors, which might be desirable from a scientific point of view, might not work in science communication and might lead to confusion and disorientation. Indeed, scientists may need ever more breakthrough or even catastrophe metaphors in situations where they have to compete with other research groups and where they have to justify their work in public (Weingart 2002). Weingart (1998) has studied the ever-closer intermingling of science and the media, which is demonstrated in the pre-publication of scientific results and the relationship between media prominence and scientific reputation. Scientists are keen to publish their results before other, competing research teams because this means they will be better placed to apply for patents and to gain public recognition in the form of prizes and awards.

The perils of such a race between scientific competitors framed by the science as *journey* metaphor were highlighted through the Korean stem cell scandal, where Korean researchers were in a race, mainly with British researchers, to achieve advances or breakthroughs in therapeutic cloning. This has led to calls for the discourse of scientific hubris based on breakthrough metaphors to be abandoned and replaced with a discourse of humility based on highlighting the progress of science as one of incremental steps, as trial-and-error and as a process rather than a series of spectacular products (see Rick Borchelt, reported by Wolvaardt 2007). The scandal also demonstrated, yet again, how scientific reputation is increasingly related to scientists' public appearances (Dunwoody 1999). This, of course, may turn into a vice in the case of the race for pre-publication.

The public debate about techno-scientific projects such as genetics and modern biotechnology is based on a complex intertwining of different discourses of science, research and development, markets, politics and economics. Here, metaphors are apt tools for communication. Some metaphors based on the scenarios of science as a *journey/race* are more precarious than others for framing scientific advances. Some discourse metaphors, such as the *book of life*, are more apt than others for framing scientific innovation, as they carry positive and well entrenched cultural associations from one context to another, and will not easily be replaced by any alternative metaphor. At the same time, certain discourse metaphors may have accelerated what Väliverronen (1993) calls a 'change from "publish or perish" to "appear in public or perish" science', a situation where scientists need to sell their research and themselves in public (Nelkin 1995). And, here – as formulated by Aristotle in his *Poetics* – 'the greatest thing by far is to be a master of metaphor'. But beware of its pitfalls.

Acknowledgements

Brigitte Nerlich would like to acknowledge funding by the Leverhulme Trust, which supports the Institute for the Study of Science, Technology and Society at the University of Nottingham, UK. She would also like to thank Andrew Balmer for his helpful comments on an earlier draft of this chapter. Iina Hellsten gratefully acknowledges the EU for partially funding this paper under the CREEN project, contract no. FP6-2003-NEST-path-012864. She would like to thank Loet Leydesdorff and Sally Wyatt for helpful comments on an earlier draft of the chapter.

Notes

1 The metaphor that frames genetically modified food as 'Frankenfood', for example (Hellsten 2003) could only be created and understood in a society in which Mary Shelley's famous novel of 1831 was a 'cultural commonplace' (Black 1962), and only after that specific novel had been written. However, once established, this type of metaphor could spread to various other target domains (in this case, related hyponymically to the original one), such as Frankencrops, Frankenplants, Frankenrice, Frankenfish and Frankensalmon. Information scientists have developed ways to trace this automatically (Thelwall and Price 2006).

2 Later, this number was reduced even further to between 20,000 and 25,000 genes – about the same amount as a fruit fly has (*How Many Genes Are in the Human Genome?*, www.ornl.gov/sci/techresources/Human_Genome/faq/genenumber.shtml).

3 '[Erwin] Schrödinger compared the genetic material to an aperyodic crystal where information could be coded in a linear array. Years later Francois Jacob discussed with Roman Jakobson the similarities of language and heredity as systems built on elements without meaning. Later on, Neils Jerne in his Nobel Prize lecture talked about the "generative grammar" of the immune system' (Berwick et al. 1998).

Suggested further reading

Ceccarelli, L. (2004) 'Neither confusing cacophony nor culinary complements: a case study of mixed metaphors for genomic science', *Written Communication*, 21: 92–105.

Charteris-Black, J. (2004) *Corpus Approaches to Critical Metaphor Analysis*, New York: Palgrave Macmillan.

Montuschi, E. (1995) 'What is wrong with talking of metaphors in science?', in Radman, Z. (ed.) *From a Metaphorical Point of View: A Multidisciplinary Approach to Cognitive Content of Metaphor*, Berlin, New York: de Gruyter, 309–27.

White, R. (1996) *The Structure of Metaphor. The Way Language of Metaphor Works*, Cambridge: Blackwell.

Other references

Anton, T. and McCourt, R. (1995) *The New Science Journalists*, New York: Ballantine Books.

Avise, J. C. (2001) 'Evolving genomic metaphors: a new look at the language of DNA', *Science* 294: 86–7.

Benford, R. and Snow, D. (2000) 'Framing processes and social movements: an overview and assessment', *Annual Review of Sociology*, 26: 611–39.

Berwick, R. C., Collado-Vides, J. and Gaasterland, T. (1998) *The Linguistics of Biology and the Biology of Language workshop*, March 23–27 1998, Morelos, Mexico.

Black, M. (1962) *Models and Metaphors*, Ithaca/London: Cornell University Press.

Bogner, A. and Menz, W. (2006) 'Science crime. The Korean cloning scandal and the role of ethics', *Science and Public Policy*, 33: 585–612.

Bono, J. (1990) 'Science, discourse and literature. The role/rule of metaphor in science', in Peterfreund, S. (ed.) *Literature and Science: Theory and Practice*, Boston, MA: Unwin Hyman, 59–89.

Bostanci, A. (2006) 'Two drafts, one genome? Human diversity and human genome research', *Science as Culture*, 15: 183–98.

Brown, N. and Webster, A. (2004) *New Medical Technologies and Society: Reordering Life*, Cambridge: Polity Press.

Bucchi, M. (1998) *Science and the Media: Alternative Routes in Scientific Communication*, London: Routledge.

—— (2004) 'Can genetics help us re-think communication? Public communication of science as "double helix" ', *New Genetics & Society*, 23: 269–83.

Burke, K. (1966) *Language as Symbolic Action*, Berkeley, CA: University of California Press.

Butler, D. (2001) 'Publication of human genome sparks fresh sequence debate', *Nature*, 409: 747–8.

Calsamiglia, H. and van Dijk, T. A. (2004) 'Popularization discourse and knowledge about the genome', *Discourse & Society*, 15: 369–89.

Ceccarelli, L. (2004) 'Neither confusing cacophony nor culinary complements: a case study of mixed metaphors for genomic science', *Written Communication*, 21: 92–105.

Charteris-Black, J. (2004) *Corpus Approaches to Critical Metaphor Analysis*, New York: Palgrave Macmillan.

Chiappe, D. L. (1998) 'Similarity, relevance, and the comparison process', *Metaphor and Symbol*, 13: 17–30.

Cohen, F. (1994) *The Scientific Revolution. Historiographical Inquiry*, Chicago, IL: University of Chicago Press.

Collins, F., Morgan, M. and Patrinos, A. (2003) 'The Human Genome Project: lessons from large-scale biology', *Science*, 300: 286–90.

Condit, C. (1999) *The Meanings of the Gene. Public Debates about Human Heredity*, Madison, WI: University of Wisconsin Press.

Cyranoski, D. (2006) 'Rise and fall: why did Hwang fake his data, how did he get away with it, and how was the fraud found out?', *Nature News*, 11 January 2006 (online).

Dunwoody, S. (1999) 'Scientists, journalists, and the meaning of uncertainty', in Friedman, S., Dunwoody, S. and Rogers, C. (eds) *Communicating Uncertainty. Media Coverage of New and Controversial Science*, London: Lawrence Erlbaum, 59–79.

Dupré, J. (2007) 'The selfish gene meets the friendly germ', keynote speech at the 22nd Regional Conference on the History and Philosophy of Science, 13–15 April 2007, Boulder, CO: University of Colorado.

Durant, J., Evans, G. and Thomas, G. (1992) 'Public understanding of science in Britain: the role of medicine in the popular representation of science', *Public Understanding of Science*, 1: 161–82.

Einseidel, E. F. (2006) 'The challenges of translating genomic knowledge', *Clinical Genetics*, 70: 433–7.

Franzen, M., Rödder, S. and Weingart, P. (2007) 'Fraud: causes and culprits as perceived by science and the media. Institutional changes, rather than individual motivations, encourage misconduct', *EMBO Reports*, 8: 3–7.

Gamson, W. and Modigliani, A. (1989) 'Media discourse and public opinion on nuclear power: a constructionist approach', *American Journal of Sociology*, 95: 1–37.

Gaskell, G. and Bauer, M. (2001) (eds) *Biotechnology 1996–2000: The Years of Controversy*, London: Science Museum.

Gogorosi, E. (2005) 'Untying the Gordian knot of creation: metaphors for the Human Genome Project in Greek newspapers', *New Genetics and Society*, 24: 299–315.

Gottweis, H. and Triendl, R. (2006) 'South Korean policy failure and the Hwang debacle', *Nature Biotechnology*, 24: 141–3.

Gross, L. (2001) *Book of Life or Signpost to a New Biology? Mapping The Human Genome Brings Surprises and New Challenges*, note on Exploratorium website: www.exploratorium.edu/aaas-2001/dispatches/bookoflife.html

Hallyn, F. (2000) 'Atoms and letters', in Hallyn, F. (ed.) *Metaphor and Analogy in the Sciences*, Origins, Studies in the Sources of Scientific Creativity, Vol. 1, Dordrecht: Kluwer Academic, 53–69.

Hawkes, T. (1972) *Metaphor*, London and New York: Routledge.

Hellsten, I. (2000) 'Dolly: scientific breakthrough or Frankenstein's Monster? Journalistic and scientific metaphors of cloning', *Metaphor and Symbol*, 15: 213–21.

—— (2001) 'Opening the Book of Life: politics of metaphors and the human genome', in Kivikuru, U. and Savolainen, T. (eds) *The Politics of Public Issues*, Publication No. 5, Helsinki: University of Helsinki, Department of Communication, 179–94.

—— (2002) 'The politics of metaphor: biotechnology and biodiversity in the media', Dissertation, Tampere, Finland: Department of Journalism and Mass Communication, University of Tampere.

—— (2003) 'Focus on metaphors: the case of "Frankenfood" on the web', *Journal of Computer-Mediated Communication*, 8(4).

Henderson, L. and Kitzinger, J. (2007) 'Orchestrating a science "event": the case of the Human Genome Project', *New Genetics and Society*, 26: 65–83.

Hesse, M. (1966) *Models and Analogies in Science*, Notre Dame, IN: University of Notre Dame Press.

—— (1988) 'The cognitive claims of metaphor', *Journal of Speculative Philosophy*, 11: 1–16.

Jacob, F. and Monod, J. (1961) 'Genetic regulatory mechanisms in the synthesis of proteins', *Journal of Molecular Biology*, 3: 318–56.

Kay, L. (2000) *Who Wrote the Book of Life: A History of the Genetic Code*, Stanford, CA: Stanford University Press.

Keller, E. F. (1995) *Refiguring Life: Metaphors of Twentieth-Century Biology*, New York: Columbia University Press.

Keller, E. F. (2000) *The Century of the Gene*, Cambridge, MA: Harvard University Press.

Kevles, L. and Hood, L. (eds) (1993) *The Code of Codes. Scientific and Social Issues in the Human Genome Project*, Cambridge, MA: Harvard University Press.

Knorr, K. (1981) 'The scientist as an analogical reasoner: a critique of the metaphor theory of innovation', in Knorr, K., Krohn, R. and Whitley, R. (eds) *The Social Process of Scientific Investigation*, Dordrecht, Boston and London: D. Reidel, 25–52.

Knudsen, S. (2005) 'Communicating novel and conventional scientific metaphors: a study of the development of the metaphor of genetic code', *Public Understanding of Science*, 14: 373–92.

Lakoff, G. and Johnson, M. (1980) *Metaphors We Live By*, Chicago, IL: Chicago University Press.

Lander, E. S. et al. (2001) 'Initial sequencing and analysis of the human genome', *Nature*, 409: 860–921.

Leatherdale, W. H. (1974) *The Role of Analogy, Model, and Metaphor in Science*, Amsterdam: North-Holland.

Leiss, W. (1985) 'Technology and degeneration: the sublime machine', in Chamberline, J. E. and Gilman, S. (eds) *Degeneration: The Dark Side of Progress*, New York: Columbia University Press.

Maasen, S. and Weingart, P. (2000) *Metaphors and the Dynamics of Knowledge*, London, New York: Routledge.

McInerney, C., Bird, N. and Nucci, M. (2004) 'The flow of scientific knowledge from the lab to the lay public. The case of genetically modified food', *Science Communication*, 26: 44–74.

Mio, J. S. and Katz, A. (eds) (1996) *Metaphor: Implications and Applications*, Mahwah, NJ: Erlbaum.

Mithen, S. (1998) *The Prehistory of the Mind: A Search for the Origins of Art, Religion and Science*, London: Phoenix/Orion.

Montuschi, E. (1995) 'What is wrong with talking of metaphors in science?', in Radman, Z. (ed.) *From a Metaphorical Point of View: A Multidisciplinary Approach to Cognitive Content of Metaphor*, Berlin and New York: de Gruyter, 309–27.

Moscovici, S. (1984) 'The phenomenon of social representations', in Farr, R. and Moscovici, S. (eds) *Social Representations*, New York: Cambridge University Press, 85–100.

Nelkin, D. (1994) 'Promotional metaphors and their popular appeal', *Public Understanding of Science*, 3: 25–31.

Nelkin, D. (1995) *Selling Science: How the Press Covers Science and Technology*, New York: W. H. Freeman (first published 1987).

Nelkin, D. and Lindee, S. (1995) *The DNA Mystique. The Gene as a Cultural Icon*, New York: W. H. Freeman.

Nerlich, B. and Dingwall, R. (2003) 'Deciphering the human genome: the semantic and ideological foundations of genetic and genomic discourse', in Dirven, R., Frank R. and Pütz, M. (eds) *Cognitive Models in Language and Thought: Ideology, Metaphors and Meanings*, Berlin: Mouton de Gruyter, 395–428.

Nerlich, B. and Hellsten, I. (2004) 'Genomics: shifts in metaphorical landscape between 2001 and 2003', *New Genetics and Society*, 23: 255–68.

Nerlich, B. and Kidd, K. (2005) Introduction to special issue on 'The New Genetics: Linguistic and Literary Metaphors and the Social Construction of the Genome', *New Genetics and Society*, 24: 263–6.

Nerlich, B., Clarke, D. D. and Dingwall, R. (2000) 'Clones and crops: the use of stock characters and word play in two debates about bioengineering', *Metaphor and Symbol*, 15: 223–40.

Nerlich, B., Clarke, D. D. and Dingwall, R. (2001) 'Fictions, fantasies, and fears: the literary foundations of the cloning debate', *Journal of Literary Semantics*, 30: 37–52.

Nerlich, B., Dingwall, R. and Clarke, D. D. (2002) 'The Book of Life: how the human genome project was revealed to the public', *Health: An Interdisciplinary Journal for the Social Study of Health, Illness and Medicine*, 6: 445–69.

Noble, D. (2006) *The Music of Life: Biology Beyond the Genome*, Oxford: Oxford University Press.

Petersen, A. (2005) 'The metaphors of risk: biotechnology in the news', *Health, Risk & Society*, 7: 203–8.

Rheinberger, H.-J. and Gaudillière, J.-P. (eds) (2004) *From Molecular Genetics to Genomics: The Mapping Cultures of Twentieth-Century Genetics*, Studies in the History of Science, Technology and Medicine, London: Routledge.

Richards, I. A. (1989) 'The philosophy of rhetoric', in Johnson, M. (ed.) *Philosophical Perspectives on Metaphor*, Minneapolis, MN: University of Minnesota Press, 48–62 (originally published 1936).

Rosner, M. and Johnson, T. R. (1995) 'Telling stories: metaphors of the Human Genome Project', *Hypatia*, 10: 104–29.

Schön, D. A. (1993) 'Generative metaphor: a perspective on problem-setting in social policy', in Ortony, A. (ed.) *Metaphor and Thought*, 2nd edn, Cambridge: Cambridge University Press.

Star, S. L. and Griesemer, J. (1989) 'Institutional ecology, "translations", and boundary objects: amateurs and professionals of Berkeley's Museum of Vertebrate Zoology, 1907–39', *Social Studies of Science*, 19: 387–402.

Thelwall, M. and Price, E. (2006) 'Language evolution and the spread of ideas: a procedure for identifying emergent hybrid word family members', *Journal of the American Society for Information Science and Technology*, 57: 1326–37.

Turney, J. (2005) 'The sociable gene', *EMBO Reports*, 6: 808–10.

Tyers, M. and Mann, M. (2003) 'From genomics to proteomics', *Nature*, 422: 193–97.

Ullmann, S. (1962) *Semantics: An Introduction to the Science of Meaning*, Oxford: Blackwell.

Väliverronen, E. (1993) 'Science and the mass media: changing relations', *Science Studies*, 6: 23–31.

—— (1998) 'Biodiversity and the power of metaphor in environmental issues', *Science Studies*, 11: 19–34.

Van Dijck, J. (1998) *Imagenation. Popular Images of Genetics*, New York: New York University Press.

Venter, C. et al. (2001) 'The sequence of the human genome', *Science*, 291: 1304–51.

Wagner, W. and Hayes N. (2005) *Everyday Discourse and Common-Sense – The Theory of Social Representations*, Basingstoke: Palgrave Macmillan.

Watson, J. D. and Crick, F. (1953) 'Molecular structure of nucleic acids: a structure for deoxyribose nucleic acids', *Nature*, 171: 737–8.

Weingart, P. (1998) 'Science and the media', *Research Policy*, 27: 869–79.

—— (2002) 'Kassandrarufe und Klimawandel', *Gegenworte*, 10: 21–5.

White, R. (1996) *The Structure of Metaphor. The Way Language of Metaphor Works*, Cambridge: Blackwell.

White House (2000) *President Clinton Announces the Completion of the First Survey of the Entire Human Genome: Hails Public and Private Efforts Leading to this Historic Achievement*, Washington, DC: White House, Office of the Press Secretary. www.ornl.gov/sci/techresources/Human_Genome/project/clinton1.shtml

Wolvaardt, E. (2007) 'How journalism can hide the truth about science', *SciDevNet*, 5 January.

Zinken, J., Hellsten, I. and Nerlich, B. (2003) 'What is "cultural" about conceptual metaphors?', *International Journal of Communication*, 13: 5–29.

Zinken, J., Hellsten, I. and Nerlich, B. (2008) 'Discourse metaphors', in Frank, R. Dirven, R., Ziemke, T. and Zlatev, J. (eds) *Body, Language and Mind, Vol. 2: Interrelations between Biology, Linguistics and Culture*, Amsterdam: John Benjamins.

8

Survey research on public understanding of science

Martin W. Bauer

The term 'public understanding of science' (PUS) has a dual meaning. First, it covers a wide field of activities that aim at bringing science closer to the people and promoting PUS in the tradition of a public rhetoric of science (see Fuller 2001 for the idea; OECD 1997; Miller et al. 2002 for attempted inventories of such initiatives). Second, it refers to social research that investigates, using empirical methods, what the public's understanding of science might be and how this might vary across time and context. This includes the conceptual analysis of the term 'understanding'.

This chapter concentrates on the latter, and focuses on the discussions raised by research using large-scale nationally and internationally representative sample surveys that ask people lists of prepared standard questions from a questionnaire. I review the changing research agenda by typifying three 'paradigms' of PUS research according to the questions they raised, the interventions they supported and the criticisms they attracted. The chapter ends with a short outlook on the potential for future research and a brief afterthought on the 'public deficit concept' and survey-based investigations. This review expands on previous reviews of the field (Pion and Lipsey 1981; Wynne 1995; Miller 2004).

Survey research in PUS

Table 8.1 lists the main surveys of PUS among adult populations since the 1970s, typically with nationally representative samples of 1000+ interviews in any one context. The list is not a comprehensive inventory, but shows the best-known surveys of scientific literacy, public interest in and attitudes to science that are partially comparable: the US National Science Foundation's indicator series since 1979 (e.g. NSF 2002); the Eurobarometer survey series since 1978, covering initially eight and recently 32 European countries; the national UK (ESRC, OST, Wellcome Trust) and French series (CEVIPOV; Boy 1989, 1993) reaching back to the mid-1980s and early 1970s respectively. The earliest survey of this kind seems to date from 1957, just before Sputnik shocked the Western world (Withey 1959). Later efforts came

Table 8.1 50 years of country surveys of public understanding of science

Year	*UK*	*France*	*Italy*	*EU*	*Bulgaria*	*US*	*Canada*	*New Zealand*	*Japan*	*India*	*China*	*Malaysia*	*Argentina*	*Brazil*
1957						Michigan								
1970														
1971														
1972		CEVIPOV				Harvard								
1973														
1974														
1975														
1976														
1977	EB7	EB7		EB7										
1978	EB10a	EB10a		EB10a										
1979						NSF								
1980														
1981														
1982		CEVIPOV												
1983						NSF								
1984														
1985						NSF								
1986	MORI													
1987														CNPq
1988	ESRC	CEVIPOV				NSF								
1989	EB31	EB31		EB31			MST							

***Table 8.1* (continued)**

Year	*UK*	*France*	*Italy*	*EU*	*Bulgaria*	*US*	*Canada*	*New Zealand*	*Japan*	*India*	*China*	*Malaysia*	*Argentina*	*Brazil*
1990						NSF								
1991									NISTEP		MST			
1992	EB38.1	EB38.1		EB38.1	BAS-IS	NSF								
1993														
1994														
1995						NSF					CAST			
1996	OST/Well				BAS-IS									
1997						NSF		MST			CAST			
1998														
1999						NSF								
2000	OST/Well											STIC		
2001	EB55.2	EB55.2		EB55.2	EB				NISTEP		CAST			
2002														
2003			OBSERVA								CAST		RICYT	FAPESP
2004	OST									NCAER				
2005	EB63.1	EB63.1	OBSERVA	EB63.1	EB63.1									
2006			OBSERVA			NSF-GSS								MST

Note:
For abbreviations see Box 8.1.

from Canada (MST), New Zealand (MST), Malaysia (STIC), India (NCAER), China (MST, CAST), Japan (NISTEP), Italy (Observa), Brazil (CNPq, FAPESP) and Latin America in general (RICYT). The agencies that commissioned these surveys are, for the most part, government bodies promoting science and technology (see Box 8.1).

Many relevant surveys that are related to specific and often controversial developments are not listed here. For example, over the years the International Social Survey (ISS) consortium, Eurobarometer, and national and international polling companies have conducted attitude surveys on specific topics such as nuclear energy, computer and information technology, the environment, biotechnology and genetic engineering, and, most recently, nanotechnology. Also the 'busy industry' of risk analysis has collected numerous samples of data on 'risk perceptions' of various technologies. Then, there are many surveys of adolescents' understanding of science, culminating in the global assessment of scientific literacy at school age by PISA in 2006. These more specific cases are not included in this review.

A complete inventory of relevant national and international surveys of specific topics within science and technology remains to be carried out. However, the sizable corpus of nationally and internationally comparable survey data accumulated over the past 40 years offers opportunities for secondary analysis, dynamic modelling and global comparisons that call for a renewed research effort.

Box 8.1 National and international institutions that have sponsored or conducted PUS surveys

BAS-IS Bulgarian Academy of Sciences, Institute of Sociology, Sofia
CAST China Association for Science and Technology
CEVIPOF Centre d'Etude de la Vie Politique Française, Sciences Po, Paris
CNPq Conselho Nacional de Desenvolvimento Científico e Tecnológico (Brazilian National Council for Scientific and Technological Development)
EB, Eurobarometer DG-12, later DG Research, Brussels
ESRC Economic and Social Research Council, UK
FAPESP State of São Paulo Research Foundation, Brazil
ISS International Social Survey consortium
MORI British public opinion research company
NCAER National Centre for Applied Economic Research, Delhi, India
NISTEP National Institute of Science and Technology Policy, Japan
NISTAD National Institute of Science, Technology and Development Studies, India
NSF US National Science Foundation, Washington, DC, USA
MST Ministry of Science and Technology (Canada, China, Brazil)
Observa – Science in Society Italian non-profit centre for science in society research
OST Office of Science and Technology, UK
PISA Programme for International Student Assessment, Organisation for Economic Co-operation and Development, Paris
RICYT Ibero-American Network of Science and Technology Indicators
STIC Strategic Thrust Implementation Committee, Malaysia
Wellcome Trust Research foundation, London, UK

Paradigms of researching PUS

Over the past 20 years, PUS has spawned a field of enquiry that engages, to a greater or lesser degree, sociology, psychology, history, communication studies and science policy analysis. It remains somewhat marginal, but vigorous in its output. Table 8.2 gives a schematic overview of three paradigms of research into PUS. Each paradigm has its prime time, and is characterised by a diagnosis of the problem that science faces in its relationship with the public. A key feature of each paradigm is the attribution of a deficit, either to the public or to science. Each paradigm pursues particular research questions through survey research, and offers particular solutions to the diagnosed deficit problems.

Scientific literacy (1960s to mid-1980s)

Scientific literacy builds on two ideas: first, science education is essentially part of the secular drive for basic literacy in reading, writing and numeracy; second, science literacy is a necessary part of civic competence. In a democracy, people take part in political decisions in one way or the other, either directly through voting, or indirectly via expressions of public opinion. However, the political actor is only effective if she or he is also familiar with the political process (Althaus 1998). The assumption is that scientific as well as political ignorance breeds alienation and extremism, hence the quest for 'civic scientific literacy' (Miller 1998). These ideas highlight the dangers of a cognitive deficit, and call for more and better science education through the life cycle. However, they also play to technocratic attitudes among elites: the public is *de facto* ignorant and therefore disqualified from participating in policy decisions.

An influential definition of 'science literacy' was proposed by Jon D. Miller (1983, 1992). It includes four elements: (a) knowledge of basic textbook facts of science; (b) an understanding of methods, such as probability reasoning and experimental design; (c) an appreciation of the positive outcomes of science and technology for society; and (d) the rejection of 'superstition'. Miller developed these literacy indicators from earlier work (Withey 1959; see also the review by Etzioni and Nunn 1976) supported

Table 8.2 Paradigms, problems and solutions

Period	*Attribution Diagnosis*	*Strategy Research*
Science literacy 1960s to 1985	Public deficit Knowledge	Measurement of literacy Education
Public understanding 1985–95	Public deficit Attitudes	Knowledge × attitude Attitude change Education Public relations
Science and society 1995 to present	Trust deficit Expert deficit Notions of the public Crisis of confidence	Participation Deliberation 'Angels', mediators Impact evaluation

Source: Bauer et al. 2007

by the US National Science Foundation (NSF). Between 1979 and 2001 the NSF undertook a bi-annual audit of scientific literacy in the USA with representative surveys of the adult population. Similar efforts, but less regular, came in Europe in the 1980s, and since then elsewhere (Table 8.1).

The research agenda

Knowledge is the key problem of this paradigm, and is measured by quiz-like items (see famous examples in Box 8.2). Respondents are asked to decide whether a statement of a scientific fact was true or false. Some of these items became notorious, travelled far and hit the news headlines, not least because of their scandal value.

Respondents score a point for every correct answer. It is a problem to find short and unambiguous statements that have one correct and authoritative answer, and to balance items that are easy and difficult and from different fields of science. The responses to these items, often between 10 or 20 in a given survey, must be correlated to form a reliable index of knowledge. Research involves the construction of such items and the testing of their scalar value. 'Constructing' is the right word

Box 8.2 Examples of knowledge and attitude items in literacy research

Knowledge items

'Does the Earth go around the Sun or does the Sun go around the Earth?' (*the Earth goes around the Sun*; the Sun goes around the Earth; DK).

'The centre of the Earth is very hot' (*true*, false, DK)

'Electrons are smaller than atoms' (*true*, false, DK)

'Antibiotics kill viruses as well as bacteria' (true, *false*, DK)

'The earliest humans lived at the same time as the dinosaurs' (true, *false*, DK)

Attitude items (Likert scales)

'Science and technology are making our lives healthier, easier and more comfortable' *(agree = positive)*

'The benefits of science are greater than any harmful effects' *(agree = positive)*

'We depend too much on science and not enough on faith' *(disagree = positive)*

1 strongly agree;
2 agree to some extent;
3 neither/nor;
4 disagree to some extent;
5 strongly disagree;
9 don't know (DK).

(Source: e.g. Eurobarometer 31, 1989.)

because finding and testing such items is a bit like brick laying, it has to stand up in the end. Recently, item response theory has been brought into the discussion for this purpose (Miller and Pardo 2000). As indicators, these items are only reliable in combination; any isolated single item has little significance. But, also in batteries the reliability of these scales is an issue (Pardo and Calvo 2004). However, public speakers and the mass media repeatedly pick out stand-alone items as indicators of public ignorance and reasons for public alarm. For example, the item about the Sun and the Earth (Box 8.2) has seen many citations out of context.[1]

What counts as scientific knowledge? Miller (1983) suggested two dimensions: facts and methods. This stimulated efforts to measure people's understanding of probability reasoning, experimental design, the importance of theory and hypothesis testing. Critics have argued that the essence of science is process rather than facts (Collins and Pinch 1993). Therefore topics such as theory testing, probability and uncertainty, peer review, scientific controversies, and the need to replicate experiments should be included in the assessment of literacy. Scales of methodological knowledge are more challenging to construct and, if they are employed, may still be based on very few items. An open question suggested by Withey (1959) proved useful: 'Tell me in your own words, what does it mean to study something scientifically?' Respondents' answers are coded for methodological and institutional awareness. However, the coding remains controversial: normative or descriptive (Bauer and Schoon 1993)? Process also includes awareness of scientific institutions and its procedures, what Prewitt (1983) called 'scientific savvy' and Wynne (1995) called the 'body language' of science. The latter dimension has received some attention in exploratory studies, but awaits implementation in international surveys (Bauer et al. 2000b; Sturgis and Allum 2004).

Researchers also looked at the 'don't know' (DK) responses to knowledge items, and revealed ambiguity in self-attributed ignorance. Comparing incorrect responses and DK responses suggests an index of confidence: women and some social milieus consistently prefer declaring ignorance rather than guessing, as they are less confident to opine on science (Bauer 1996). Turner and Michael (1996) described four qualitative types of admitted ignorance: embarrassment: 'I go and find a book in the library'; self-identity: 'I am not very scientific'; division of labour: 'I know somebody who knows'; smugness: 'I couldn't care less'. Sudden changes in the rates of DK responses over time might also indicate methodological issues: survey companies can alter their interviewer protocol, accept DK as an answer, or probe further. The latter reduces the rate of DKs. Such changes might reflect a change of field work contractor, as in the case of Eurobarometer, over the years.

Many countries have undertaken audits of adult scientific literacy. The US NSF has presented bi-annual 'horse race'-type rankings of different countries on literacy to answer the question: where does the USA stand? A key problem of such comparisons remains the fairness of the indicators. The existing set of knowledge items is biased with regard to the national science base. Countries tend to have a historically specialised science base, or a corps of scientific heroes from one rather than the other discipline, and literacy scores are likely to reflect this (Shukla 2005; Raza et al. 1996, 2002). Raza and colleagues' suggestions for culturally fair indicators and analysis deserve more attention than they have received.

What's to be done?

The literacy paradigm is fixated on the cognitive deficit. Interventions are seen mainly in public education. Literacy is a matter for continued education, and requires renewed attention in schools curricula and on the part of a publicly responsible mass media that is called to task (Royal Society 1985).

Critique

The critique of the literacy paradigm focuses on conceptual as well as empirical issues. Why should science knowledge qualify for special attention? What about historical, financial or legal literacy? The case for 'science literacy' needs to be made in competition with other types of literacy.

Second, is literacy a continuum or a threshold measure? Miller (1983) originally envisaged a threshold measure. To qualify as a member of the 'attentive public for science', in Miller's view, one needs to command 'some minimal level' of literacy, be interested and feel informed about science and technology, appreciate some positive outcomes, and renounce superstitions. However, the definition of this 'minimal level of literacy' changed from audit to audit, and it is unclear whether the reported shifts, or for that matter their absence (Miller 2004), reflect changes in definition or in substance (Beveridge and Rudell 1998).

Third, critics argued that indicators of 'textbook knowledge' are irrelevant and empirical artefacts. Of real importance is knowledge in context, which emerges from local controversies and people's concerns (Ziman 1991; Irwin and Wynne 1996). However, what accounts for the consistent correlations between measures of literacy, attitudes and sociodemographic variables? We must recognise a certain intellectual failure to engage with these robust results.

Fourth, there is the question of 'superstitions'. Does belief in astrology disqualify a member of the public from being scientifically literate, as Miller (1983) suggested? The coexistence of superstition and scientific literacy is an empirical matter. Astrology and scientific practice serve different functions in life. To make the 'rejection of astrology' a criterion of literacy bars us from understanding the tolerance or intolerance between science and astrology in everyday life, which is a cultural variable (Boy and Michelat 1986; Bauer and Durant 1997).

Knowledge items can be controversial in substance. So, for example, physicists might point out that whether 'electrons are smaller than atoms' cannot be determined in general, but depends on circumstances. Thus the authoritative answer to this question is shakier than is admitted, though a generally correct textbook answer may be deemed acceptable. More problematic is the statement 'The earliest humans lived at the same time as the dinosaurs', which, according to biology textbooks, is 'false'. This item captures, in the USA in particular, a debate over evolutionary theory that (by no means globally) clashes with fundamentalist religious culture (Miller et al. 2006). It is thus not entirely clear whether this item is to be treated as an indicator of science literacy or of cultural values; this will depend on the context.

Finally, it is suggested that the concern with literacy is correlated with the crisis of legitimacy of 'big science'. To overcome this crisis by literacy assumes a fundamental gap in the operations of literate scientists and an illiterate public, for which there is little evidence beyond elitist prejudice. Furthermore, if Roger Bacon's notion of 'knowledge = power' holds, any attempt to share knowledge without simultaneous public empowerment will create alienation rather than rapprochement between science and the public. Literacy is therefore the wrong answer to what many see as a crisis of legitimacy and trust (Roqueplo 1974; Fuller 2000).

Public understanding of science (1985 to mid-1990s)

In this period, new concerns emerged under the title 'public understanding of science'.[2] In the UK this was marked by an internationally influential report of the Royal Society of London (Royal Society 1985). PUS inherited the notion of a public deficit, but now it was the attitudinal deficit that was in the foreground (Bodmer 1987). The public was seen to be not sufficiently positive about science and technology, too sceptical or even outright anti-science. This was of major concern to scientific institutions such as the Royal Society. Old and new good reasons for the public appreciation of science were put forward: it is important for making informed consumer choices; it enhances the competitiveness of industry and commerce; and it is part of national tradition and culture (Thomas and Durant 1987; Gregory and Miller 1998; Felt 2000). The Royal Society famously assumed that more public knowledge will 'cause' more positive attitudes. Hence, the axiom of PUS became: 'the more they know, the more they love it'.

Research agenda

The evaluative appreciation of science is measured mostly by Likert-type attitude scales. Respondents agree or disagree with evaluative statements, and thereby express their positive or negative attitude towards science (see Box 8.2 for some famous examples). Some statements, in order to assess a positive attitude, require the respondent to disagree, others to agree, depending on the formulation. This mixed tactic avoids the acquiescence response bias – the general tendency to agree to such questions in the artificial context of survey interviews. Another related issue is how to deal with 'neither/nor' and DK options. Not offering a 'neither/nor' may increase the variance in the data, but forces people into positions that they do not hold. This would leave no space to express real ambivalence, genuinely motivated abstention of judgement, or the absence of opinion. There must be space for the 'idiot' (in ancient Greece, the one who does not have an opinion in public; on the recent revaluation of the 'idiot' see Lezaun and Soneryd 2006).

Research on attitudes to science is concerned with 'acquiescence response bias', the construction of reliable scales, the multidimensional structure of attitudes (Pardo and Calvo 2002), the relationship between general and specific attitudes (Daamen and vanderLans 1995), context effects of previous questions (Gaskell et al. 1993), the comparative effects of web-based or telephone interviewing on the response variance (Fricker et al. 2005) and, most importantly, the relationship between knowledge and attitudes. The concern for literacy carried over into PUS, as knowledge measures are needed to test the expectation 'the more they know, the more they love it'. However, the emphasis shifts from a threshold measure to that of a continuum of knowledge. One is either literate or not, but more or less knowledgeable (Durant et al. 1989).

A potential gender gap in attitudes to science attracts attention and concern. Breakwell and Robertson (2001) found that British girls are less inclined towards science than boys, and this gap is unchanged between 1987 and 1998. Sturgis and Allum (2001) highlight the importance of a knowledge gap between men and women in explaining this attitude gap, when controlling for other factors. Equally, for Switzerland in 2000, Crettaz von Roten (2004) shows that gender does not explain attitudinal differences between men and women; it is science literacy and general education that makes the difference.

The second PUS paradigm extended the range of concepts, methods and data. Mass media monitoring, in particular of newspapers, is cost-effective and easily extended backward in time and updated into the present. The changing salience and the framing of science in the mass media offer alternative indicators of science in public (Bauer 2000). Such analyses reveal long-term cycles and trends, such as the return to the medicalisation of science news, over the past 100 years (LaFollette 1990; Bauer 1998; Bucchi and Mazzolini 2003; Bauer et al. 2006).

What's to be done?

The practical interventions based on the PUS paradigm might be divided into a rationalist and a realist agenda. Both agree with the diagnosis of an attitudinal deficit: the public is insufficiently infatuated with science and technology, but disagrees on what to do about it. For the rationalist, public attitudes are a product of information processing with a cognitive–rational core. Hence negative attitudes towards science – or unreasonable risk perceptions, as was the fashionable concept of the 1990s – are caused by insufficient information, or they are based on heuristics, such as availability or small sample evidence, that bias the public's judgement of science and technology. It is assumed that, if people had all the information and operated without these heuristics, they would display more positive judgements of scientific developments. Thus they would agree with experts, who do not succumb to these biases as easily as the public does. People need more information and training on how to avoid faulty information processing. The battle for the public is thus a battle for rational minds with the weapons of information and training in probability and statistics.

For the realist, attitudes express emotional relations with the world. Realists work the emotions and appeal to people's desires and gut reactions, and thus follow the logic of advertising and propaganda. In what is seen as the battle for the hearts of the public, the key question is: how can we make science 'sexy'? The 'consumer' public is to be seduced rather than rationally persuaded. According to this logic, there is little difference between science and washing powder (Michael 1998). This agenda includes research on market segmentation, consumer profiling, and targeted campaigning in different segments of the public. British science consumers are divided into six groups with different demographic profiles (OST 2000): confident believers, technophiles, supporters, concerned, the 'not sure', and the 'not for me'. A similar segmentation was developed in a Portuguese study (Costa et al. 2002).

Critique

Attitudes need to be understood against a background of representations. Representations become visible when everyday common sense is challenged by novelty; they familiarise the unfamiliar (Farr 1993; Wagner 2007). This outlook shifts research from rank-ordering people by attitudes, to characterising their representations of science as a function of different life contexts (Boy 1989; Bauer and Schoon 1993; Durant et al. 1992). Studying representations of science opens the door wide to other data streams, in particular qualitative enquiries (Jovchelovitch 1996; Bauer and Gaskell 1999; Wagner and Hayes 2005).

The critique of deficit models correctly highlighted the pitfalls of reifying 'knowledge' in the knowledge survey – scientific knowledge is what surveys measure – and insisted on knowledge-in-context (Ziman 1991) and on analysing how experts relate to the public (Irwin and Wynne 1996). Wynne (1993) used the term 'institutional neuroticism' to point to prejudices of scientific actors towards the public that create a self-fulfilling prophecy and a vicious circle: the public, cognitively and emotionally deficient, cannot be trusted. However, this mistrust by scientific actors will be paid back in kind by public mistrust. Negative public attitudes then confirm the assumptions of scientists: the public is not to be trusted. This circularity of the 'institutional unconscious' calls for 'soul searching' – reflexivity among scientific actors, and even endorsement of a post-modern epistemology of a plurality of knowledge centres.

The empirical critique of the paradigm focuses on the relationships between interest, attitudes and knowledge. The correlation between knowledge and attitudes becomes a focus of research (Evans and Durant 1989; Einsiedel 1994; Durant et al. 2000). The results remained inconclusive until recently (Allum et al. 2008): overall, large-scale surveys show a small positive correlation of knowledge and positive attitudes, but they also show larger variance among the knowledgeable, and this correlation is variable. However, on controversial science topics, the correlation approaches zero. Thus not all informed citizens are also enthusiastic about all science and technology; for some developments, in particular controversial ones, 'familiarity can breed contempt'. In hindsight, it is surprising that anybody ever expected this to be different.

The concept and measurement of attitudes is the traditional remit of social psychology (Eagly and Chaiken 1993). In classical theory, cognitive elaboration is not a factor of positive attitudes, but is an indicator of their quality: knowledge fortifies the attitude of resisting influence and makes it more predictive of behaviour, whatever its direction (Pomerantz et al. 1995). What emerges is that knowledge and information matter, but not in the way common sense assumes. It is not normal that better-informed citizens have more positive attitudes, or that poorly informed and well informed citizens equally stick to their views.

Many PUS surveys also measure people's interest in science. Eurobarometer surveys suggest that self-reported interest is falling over time, while knowledge is increasing (Miller et al. 2002). This trend, if verified, would show that 'familiarity breeds disinterest', thus touching another naive assumption of the scientific literacy and PUS paradigms: the 'more we know, the more we are interested'.

Science in-and-of society (mid-1990s to present)

The polemical critique of the public deficit models ushered in a reversal of the attribution. The public deficit of trust is mirrored by a deficit on the part of science and technology and its representatives. The focus of attention shifts to the deficit of the scientific expert: their prejudices about the public.

Diagnosis

Evidence of negative attitudes from large-scale surveys is contextualised through focus group research and quasi-ethnographic observations, and reinterpreted as a 'crisis of confidence' (House of Lords 2000; Miller 2001). Science and technology operate in society and therefore stand relative to other sectors of society. The views of the public held by scientific experts come under scrutiny. Prejudices of the public operate in policy-making and guide communication efforts, and these alienate the public. The decline in trust of the public *vis à vis* science also indicates the revival of an enlightenment notion of a sceptical but informed public opinion (Bensaude-Vincent 2001).

What's to be done?

Writings on the new governance of science urge more public involvement and a new deal between science and society (Fuller 2000; Jasanoff 2005). For the 'science and society' paradigm, the distinction between research and intervention blurs. Many are committed to action research that rejects the separation of analysis and intervention. This agenda, academically grounded as it may be, often ends in political consultancy with a very pragmatic outlook. Explicit and implicit notions of the public, public opinion and public sphere are reported back as 'theories espoused' and 'theories-in-action' (Argyris and Schön 1978) to stimulate reflective change of mind among these scientific actors. Thus the bulk of activity under science and society fuses research and consultancy, and most of the evidence remains grey literature and informal conversation.

Advice proliferates on how to rebuild public trust by addressing its paradoxes: trust is relational; once it is on the cards, trust is already lost; trust cannot be engineered, it is granted to who deserves it (Luhmann 1979). Public participation is the Prince's way to rebuild public trust: the House of Lords (2000) report on *Science and Society* lists many forms of deliberative activity such as citizen juries, deliberative opinion polling, consensus conferencing, national debates and hearings. The virtues, experience and know-how of these exercises in public deliberation have been ordered and described (Seargent and Steel 1998; Einsiedel et al. 2001; Joss and Belluci 2002; Abels and Bora 2004; Gregory et al. 2007).

Critique

Deliberative activities are time-consuming, require know-how, and are thus increasingly outsourced to a newly forming private sector of professional 'angels'. Angels are age-old mediators, in this context not between heaven and earth, but between a disenchanted public and the institutions of science, industry and policy-making. For the utilitarian spirit of politics, the democratic ethos is not sufficient: the pertinent question is, does the deliberation process pay off? The private sector of outsourced professional 'angels' makes claims and offers 'deliberation services', often with spurious product differentiations, which require some kind of critical consumer testing.

The answer to the audit problem requires process and impact measures. Researchers therefore advocate quasi-experimental evaluations of deliberative events and suggest process and outcome indicators (Butschi and Nentwich 2002; Rowe and Frewer 2004), including indicators such as changing public literacy and attitudes.

This call for indicators of the public, and for media monitoring of the events, runs the risk of reinventing the wheel, albeit this time for a different car: to evaluate the 'angels' and their services. The re-entry of concerns for PUS by the back door of impact evaluation research of public deliberation is ironic, but unavoidable. The analysis of costs and benefits needs a metric.[3] It may be that changes in media reportage, public knowledge, interests and attitudes offer that metric to audit the moderators of public audits. After all, we live in an 'audit society' (Power 1999).

A brief afterthought on an urban myth: public deficits and survey research

Survey research on PUS has at times been hampered by a polemic association of the 'deficit concept' with survey research. This polemic is reminiscent of an older one, namely Lazarsfeld's (1941) debate with Adorno over 'administrative' and 'Critical' research, but with a curious methodological fixation.[4] As the polemic has it, the PUS survey researcher is a 'positivist' who constructs the 'public deficit' of knowledge, attitude and trust to please her sponsors in government, business or learned societies. Thus surveys serve existing powers to control their anxieties over public opinion. By contrast, the 'critical-constructivist' researcher will recognise this ideological entanglement, and will use exclusively qualitative data. The achieved reflexivity will open his sponsors to a change of mind. 'Critical' qualitative research emancipates the public from the grip of elite prejudice. Where doxa reigns, there shall be logos – so be it. Problematic in all this is not the differentiation of knowledge interests, nor the move from prejudice to enlightenment, but the exclusive identification of knowledge interest and method protocol:

> survey research = positivist = anxious, decontextual prejudice
> qualitative research = critical-constructivist = contextual insight.

These implicit equations are of unclear origin, like any urban myth, but probably arose from the influential British research programme on PUS of the late 1980s. In his review of PUS research, Brian Wynne (1995) associated survey research with anxious elite prejudice, and contrasted both with the alternative 'constructivist research':

> Large-scale social surveys of public attitudes toward and understanding of science inevitably build in certain normative assumptions about the public, about

> what is meant by science and scientific knowledge, and about understanding. They may often therefore reinforce the syndrome [anxiety among social elites about maintaining social control, p. 361], in which the public, and not science or scientific culture and institutions are problematised. . . . The survey method by its nature decontextualises knowledge and understanding and imposes assumptions.
>
> (Wynne 1995: 370)

Such formulations, together with the book that came to summarise the British research effort of the late 1980s (Irwin and Wynne 1996), framed an unhelpful polemic, without intentions to that effect.[5] But whereas Lazarsfeld invited convergence with Adorno, here the survey method is stigmatised by methodological boundary work (Gieryn 1999). The intrinsic association of sample surveys and elite anxieties and the control of public opinion is a logical fallacy, historically unfounded, and unduly restrictive of social research (Bauer et al. 2000a; Kallerud and Ramberg 2002). The abuse of survey data is clearly possible and documented by the polemic, which is an achievement to be appreciated. But the misuse of an instrument does not exhaust its potential. The intrinsic identification of data protocol and knowledge interest ignores the functional autonomy of motives and the 'interpretive flexibility' of instruments. Generally, motive and behaviour stream are flexibly linked, and this goes for knowledge interest and method protocols, and also for survey research. Ironically, 'critical' qualitative research is increasingly adopted by powerful sponsors, which absorbs much of the critical edge. After all, it is difficult to see why qualitative research should be immune to this type of incorporation. Nevertheless, we need to break with the stigma of survey research among PUS researchers and sponsors to liberate and expand the agenda for future research.

Where to go from here?

Progress in PUS research appears to be modest: few new questions, but many new discourses. None of the new discourses has made the previous ones obsolete. PUS preserves the relevance of literacy measurement and adds the attitudes. Science-and-society, while rejecting the public deficit models, cannot, once consolidated, avoid the thorny problem of auditing, and ironically reinvents the old measures of the public. However, the focus is no longer on the public's deficit, but on the performance of the mediating 'angels' who spend public money to resolve the crisis of public confidence in science and technology.

Equally promising are the Organisation for Economic Co-operation and Development (OECD) assessment cycles of educational achievement (PISA), which in 2006 focused on 'scientific literacy' and will do so again in 2015, albeit only for children of school age. These results create large, detailed databases across many countries, even beyond the OECD, and might also relaunch the discussions over adult literacy in these contexts.

This account of the recent history of PUS research both appreciated and deplored the 'deficit concept' in PUS research. In order to capitalise on the opportunities that arise from a global corpus of survey data that has accumulated over the past 40 years, we need to reopen the research agenda. We need to ignore the misguided stigma of survey research and accept it as a powerful and 'movable immobile' representation of public opinion. If this is achieved, this field of research will enter its most fertile period yet. Public understanding of science is a historical process. One-off point observations (the occasional survey or focus group) might have news and scandal value, but do not offer a valid analysis of the historical dynamic. Now it is time to be ambitious again (Bauer et al. 2007): let researchers come together with new efforts to:

- integrate the different national and international surveys as far as possible into a global database, maybe under the EU, World Bank, UNESCO, OECD or UN flag, and in collaboration with existing social science data archives;
- encourage sophisticated secondary analysis and the continued documentation of this growing database;
- construct dynamic models of PUS over time, including cohort analytical and quasi-panel models, and test these in different contexts;
- work towards global indicators of a 'culture of science' based on these surveys;
- seriously commit to and develop alternative data streams, such as mass media monitoring and longitudinal qualitative research efforts.

Notes

1 It would be an interesting study to trace the reception and use of these items in public discourse. I am not aware of any study analysing the reception of PUS surveys. This would amount to an interesting analysis of the rhetoric and pragmatics of survey research.

2 PUS was also extended to PUST to include 'T' for technology, PUSTE to include 'E' for engineering, or PUSH to include 'H' for the humanities, the latter indicating a more continental understanding of 'science' as *Wissenschaft*. The dating of these phases follows mainly the influential UK experience. In the USA, all through the 1970s the American Association for the Advancement of Science had a standing Committee on the Public Understanding of Science (Kohlstedt et al. 1999: 140ff).

3 In the UK, for example, the large-scale GM Nation Debate of 2002 cost the Government close to £1 million. A simple survey of public attitudes would cost around £50–100,000; a round of focus group sessions, £25,000. Such cost differentials have to be justified with added value.

4 I use the capital 'C' to indicate this posture of foundational critique. It is hard to imagine how a critical mind could be the privilege of only one side of this polemic. And this dichotomy is already a simplification of a more complex issue, which Habermas (1978) caught with the trichotomy of technical–instrumental, practical–normative, and critical–emancipatory interests of human knowledge.

5 A historical note: in the UK, the ESRC funded a research programme, 'Public understanding of science', in 1987–90 (Ziman 1992) with 11 different projects. I joined this effort in 1989 as a 'number-cruncher' for John Durant, then at the Science Museum in London. The later publication (Irwin and Wynne 1996), which became the summary of this programme, 'excluded' the three projects with numerical data: the survey of the British adult population (Durant et al. 1989, 1992); the survey of British children (Breakwell and Robertson 2001); and the analysis of mass media reportage of science (Hansen and Dickinson 1992). Alan Irwin recalls (personal communication, January 2007) that the book was intended to counterbalance

the overwhelming publicity that the survey of 1988 generated through its *Nature* piece (Durant et al. 1989), not as a statement against quantitative or survey research. But he admitted that an 'urban myth' might have developed in the subsequent reception, not least because the anti-survey polemic resonated with the issue of quantitative and qualitative methodology in the social sciences generally, and the myth might have served the boundary work during the competition over national and European funding streams.

Suggested further reading

Allum, N., Sturgis, P., Tabourazi, D. and Brunton-Smith, I. (2008) 'Science knowledge and attitudes across cultures: a meta-analysis', *Public Understanding of Science*, 17: 35–54.

Bauer M. W., Allum, N. and Miller, S. (2007) 'What can we learn from 25 years of PUS survey research? Liberating and expanding the agenda', *Public Understanding of Science*, 16: 79–98.

Miller, J. D. (2004) 'Public understanding of, and attitudes toward, scientific research: what we know and what we need to know', *Public Understanding of Science*, 13: 273–94.

References

Abels, G. and Bora, A. (2004) *Demokratische Technikbewertung*, Bielefeld: Transcript Verlag.

Allum, N., Sturgis, P., Tabourazi, D. and Brunton-Smith, I. (2008) 'Science knowledge and attitudes across cultures: a meta-analysis', *Public Understanding of Science*, 17: 35–54.

Althaus, S. (1998) 'Information effects in collective preferences', *American Political Science Review*, 69: 1218–23.

Argyris, C. and Schön, D. A. (1978) *Organizational Learning: A Theory of Action Perspective*, Reading, MA: Addison-Wesley.

Bauer, M. (1996) 'Socio-economic correlates of DK-responses in knowledge surveys', *Social Science Information*, 35: 39–68.

—— (1998) 'The medicalisation of science news: from the "rocket-scalpel' to the "gene-meteorite" complex', *Social Science Information*, 37: 731–51.

—— (2000) 'Science in the media as cultural indicators: contextualising survey with media analysis', in Dierkes, M. and vonGrote, C. (eds) *Between Understanding and Trust – The Public, Science and Technology*, Amsterdam: Harwood, 157–78.

Bauer, M. W. and Durant, J. (1997) 'Belief in astrology – a social-psychological analysis', *Culture and Cosmos – A Journal of the History of Astrology and Cultural Astronomy*, 1: 55–72.

Bauer, M. W. and Gaskell, G. (1999) 'Towards a paradigm for research on social representations', *Journal for the Theory of Social Behaviour*, 29: 163–86.

Bauer, M. W. and Schoon, I. (1993) 'Mapping variety in public understanding of science', *Public Understanding of Science*, 2: 141–55.

Bauer, M. W., Gaskell, G. and Allum, N. (2000a) 'Quantity, quality and knowledge interests: avoiding confusions', in Bauer, M.W and Gaskell, G. (eds) *Qualitative Researching with Text, Image and Sound – A Practical Handbook*, London: Sage, 3–18.

Bauer, M. W., Petkova, K. and Boyadjewa, P. (2000b) 'Public knowledge of and attitudes to science – alternative measures', *Science, Technology and Human Values*, 25: 30–51.

Bauer, M. W., Petkova, K., Boyadjieva, P. and Gornev, G. (2006) 'Long-term trends in the public representations of science across the iron curtain: Britain and Bulgaria, 1946–95', *Social Studies of Science*, 36: 97–129.

Bauer, M. W., Allum, N. and Miller, S. (2007) 'What can we learn from 25 years of PUS survey research? Liberating and expanding the agenda', *Public Understanding of Science*, 16: 79–98.

Bensaude-Vincent, B. (2001) 'A genealogy of the increasing gap between science and the public', *Public Understanding of Science*, 10: 99–113.

Beveridge, A. A. and Rudel, F. (1988) 'An evaluation of public attitudes toward science and technology in Science Indicators: the 1985 report', *Public Opinion Quarterly*, 52: 374–85.

Bodmer, W. (1987) 'The public understanding of science', *Science and Public Affairs*, 2: 69–90.

Boy, D. (1989) *Les Attitudes des Français a l'Egard de la Science*, Paris: CNRS, CEVIPOF.

—— (1993) 'Les Français et les techniques' in *Actes du colloque des 12 et 13 Decembre, 1991, Paris: Palais de la Decouverte*, 113–122.

Boy, D. and Michelat, G. (1986) 'Croyance aux parasciences: dimensions sociales et culturelles', *Revue Francaise de Sociologie*, 27: 175–204.

Breakwell, G. M. and Robertson, T. (2001) 'The gender gap in science attitudes, parental and peer influences: changes between 1987/88 and 1997/98', *Public Understanding of Science*, 10: 71–82.

Bucchi, M. and Mazzolini, R. (2003) 'Big science, little news: science coverage in the Italian daily press, 1946–1987', *Public Understanding of Science*, 12: 7–24.

Butschi, D. and Nentwich, M. (2002) 'The role of participatory technology assessment in the policy-making process', in Joss, S. and Bellucci, S. (eds) *Participatory Technology Assessment*, London: Centre for the Study of Democracy.

Collins, H. and Pinch, T. (1993) *The Golem. What Everyone Should Know about Science*, Cambridge: Cambridge University Press.

da Costa, A. R., Avila, P. and Mateus, S. (2002) *Publicos da Ciencia em Portugal*, Lisbon: Gravida.

Crettaz von Roten, F. (2004) 'Gender differences in attitudes towards science in Switzerland', *Public Understanding of Science*, 13: 191-9.

Daamen, D. D. L. and vanderLans, I. A. (1995) 'The changeability of public opinions about new technology: assimilation effects in attitude surveys', in Bauer, M. (ed.) *Resistance to New Technology*, Cambridge: Cambridge University Press, 81–96.

Durant, J., Evans, G. A. and Thomas, G. P. (1989) 'The public understanding of science', *Nature*, 340: 11–14.

—— (1992) 'Public understanding of science in Britain: the role of medicine in the popular representation of science', *Public Understanding of Science*, 1: 161–82.

Durant, J., Bauer, M., Midden, C., Gaskell, G. and Liakopoulos, M. (2000) 'Two cultures of public understanding of science', in Dierkes, M. and vonGrote, C. (eds) *Between Understanding and Trust: The Public, Science and Technology*, Amsterdam: Harwood Academic, 131–56.

Eagly, A. H. and Chaiken, S. (1993) *The Psychology of Attitudes*, Fort Worth, TX: Harcourt Brace College Publishers.

Einsiedel, E. (1994) 'Mental maps of science: knowledge and attitudes among Canadian adults', *International Journal of Public Opinion Research*, 6: 35–44.

Einsiedel, E., Jelsoe, E. and Breck, T. (2001) 'Publics at the technology table: the consensus conference in Denmark, Canada and Australia', *Public Understanding of Science*, 10: 83–98.

Etzioni, A. and Nunn, C. (1976) 'The public appreciation of science in contemporary America', in Holton, G. and Planpied, W. A. (eds) *Science and its Public: The Changing Relationship*, Dordrecht: Reidel, 229–243.

Evans, G. and Durant, J. (1989) 'Understanding of science in Britain and the USA', in Jowell, R., Witherspoon, S. and Brook, L. (eds) *British Social Attitudes, 6th Report*, Aldershot: Gower, 105–120.

Farr, R. M. (1993) 'Common sense, science and social representations', *Public Understanding of Science*, 2: 189–204.

Felt, U. (2000) 'Why should the public "understand" science? A historical perspective on aspects of public understanding of science', in Dierkes, M. and vonGrote, C. (eds) *Between Understanding and Trust – The Public, Science and Technology*, Amsterdam: Harwood, 7–38.

Fricker, S., Galesic, M., Tourangeau, R. and Yan, T. (2005) 'An experimental comparison of web and telephone survey', *Public Opinion Quarterly*, 69: 370–92.

Fuller, S. (2000) *The Governance of Science*, Oxford: Oxford University Press.
Fuller, S. (2001) 'Science', in *Encyclopedia of Rhetoric*, Oxford: Oxford University Press, 703–713.
Gaskell, G., Wright, D. and O'Muircheartaigh, C. (1993) 'Measuring scientific interest: the effect of knowledge questions on interest ratings', *Public Understanding of Science*, 2: 39–58.
Gieryn, T. F. (1999) *Cultural Boundary Work of Science*, Chicago, IL: University of Chicago Press.
Gregory, J. and Miller, S. (1998) *Science in Public. Communication, Culture and Credibility*, London: Plenum.
Gregory, J., Agar, J., Lock, S. and Harris, S. (2007) 'Public engagement of science in the private sector: a new form of PR?' in Bauer, M. W. and Bucchi, M. (eds) *Journalism, Science and Society: Science Communication between News and PR*, London: Routledge.
Habermas, J. (1978) 'Erkenntnis und Interesse', *Merkur*, 213: 1139–1153; later reprinted as 'Knowledge and human interest', in Habermas, J. (1978) *Science and Technology as Ideology*, London: Heinemann (first published in 1965).
Hansen, A. and Dickinson, R. (1992) 'Science coverage in the British mass media: media output and source input', *Communication*, 17: 365–77.
House of Lords (2000) *Science and Society, 3rd Report*, London: House of Lords Select Committee on Science and Technology/HMSO.
Irwin, A. and Wynne, B. (1996) *Misunderstanding Science? The Public Reconstruction of Science and Technology*, Cambridge: Cambridge University Press.
Jasanoff, S. (2005) *Designs on Nature. Science and Democracy in Europe and the United States*, Princeton, NJ: Princeton University Press.
Joss, S. and Bellucci, S. (eds) (2002) *Participatory Technology Assessment*, London: Centre for the Study of Democracy.
Jovchelovitch, S. (1996) 'In defence of representations', *Journal for the Theory of Social Behaviour*, 26: 121–36.
Kallerud, E. and Ramberg, I. (2002) 'The order of discourse in surveys of public understanding of science', *Public Understanding of Science*, 11: 213–24.
Kohlstedt, S. G., Sokal, M. M. and Lewenstein, B. (1999) *The Establishment of Science in America*, Washington, DC: American Association for the Advancement of Science.
LaFollette, M. C. (1990) *Making Science our Own: Public Images of Science, 1910–1955*, Chicago, IL: Chicago University Press.
Lazarsfeld, P. F. (1941) 'Remarks on administrative and critical communication research', *Studies in Philosophy and Social Science*, 9: 2–16.
Lezaun, J. and Soneryd, L. (2006) 'Government by Elicitation: Engaging Stakeholders or Listening to Idiots', *Discussion Paper no. 34, London: Centre for Analysis of Risk and Regulation/London School of Economics and Political Science.*
Luhmann, N. (1979) *Trust and Power*, Chichester: Wiley (German original 1968).
Michael, M. (1998) 'Between citizen and consumer: multiplying the meanings of "public understanding of science"', *Public Understanding of Science*, 7: 313–28.
Miller, J. D. (1983) 'Scientific literacy: a conceptual and empirical review', *Daedalus*, Spring: 29–48.
Miller, J. D. (1992) 'Towards a scientific understanding of the public understanding of science and technology', *Public Understanding of Science*, 1: 23–30.
—— (1998) 'The measurement of civic scientific literacy', *Public Understanding of Science*, 7: 203–24.
—— (2004) 'Public understanding of, and attitudes toward, scientific research: what we know and what we need to know', *Public Understanding of Science*, 13: 273–94.
Miller, S. (2001) 'Public understanding of science at the cross-roads', *Public Understanding of Science*, 10: 115–20.
Miller, S., Caro, P., Koulaidis, V., de Semir, V., Staveloz, W. and Vargas, R. (2002) *Benchmarking the Promotion of RTD Culture and Public Understanding of Science*, Brussels: Commission of the European Communities.

Miller, J. D., Scott and E. C. Okamoto, S. (2006) 'Public acceptance of evolution', *Science*, 313: 765–66.

NSF (2002) *Science Indicators, Chapter 7: Public Attitudes and Understanding*, Arlington, VA: National Science Foundation.

OECD (1997) *Promoting Public Understanding of Science – A Survey of OECD Country Initiatives*, GD(97)52, Paris: Organisation for Economic Co-operation and Development.

OST (2000) *Science and the Public: A Review of Science Communication and Public Attitudes to Science in Britain*, London: Office of Science and Technology/Wellcome Trust.

Pardo, R. and Calvo, F. (2002) 'Attitudes toward science among the European public: a methodological analysis', *Public Understanding of Science*, 11: 155–95.

—— (2004) 'The cognitive dimension of public perceptions of science: methodological issues', *Public Understanding of Science*, 13: 203–27.

Pion, G. M. and Lipsey, M. W. (1981) 'Public attitudes towards science and technology: what have the surveys told us?', *Public Opinion Quarterly*, 45: 303–16.

Pomerantz, E. M., Chaiken, S. and Tordesillas, R. S. (1995) 'Attitude strength and resistance processes', *Journal of Personality and Social Psychology*, 69: 408–19.

Power, M. (1999) *The Audit Society: Rituals of Verification*, Oxford: Oxford University Press.

Prewitt, K. (1983) 'Scientific literacy and democratic theory', *Daedalus*, Spring: 49–64.

Raza, G., Singh, S., Dutt, B. and Chander, J. (1996) *Confluence of Science and People's Knowledge at the Sangam*, New Delhi: NISTED.

—— (2002) 'Public, science and cultural distance', *Science Communication*, 23: 292–309.

Roqueplo, P. (1974) *Le Partage du Savoir – Science, Culture, Vulgarisation*, Paris: Seuil.

Royal Society (1985) *The Public Understanding of Science*, London: Royal Society.

Rowe, G. and Frewer, L. (2004) 'Evaluating public participation exercises: a research agenda', *Science, Technology and Human Values*, 29: 512–56.

Seargent, J. and Steele, J. (1998) *Consulting the Public – Guidelines and Good Practice*, London: Policy Studies Institute.

Shukla, R. (2005) *India Science Report – Science Education, Human Resources and Public Atttitudes towards Science and Technology*, Delhi: NCAER.

Sturgis, P. J. and Allum, N. (2001) 'Gender differences in scientific knowledge and attitudes towards science: a reply to Hayes and Tariq', *Public Understanding of Science*, 10: 427–30.

—— (2004) 'Science in society: re-evaluating the deficit model of public attitudes', *Public Understanding of Science*, 13: 55–74.

Thomas, G. P. and Durant, J. (1987) 'Why should we promote the public understanding of science?', in Shortland, M. (ed.) *Scientific Literacy Papers*, Oxford: Rewley, 1–14.

Turner, J. and Michael, M. (1996) 'What do we know about "don't knows"? Or, contexts of ignorance', *Social Science Information*, 35: 15–37.

Wagner, W. (2007) 'Vernacular science knowledge: its role in everyday life communication', *Public Understanding of Science*, 16: 7–22.

Wagner, W. and Hayes, N. (2005) *Everyday Discourse and Common Sense. The Theory of Social Representations*, London: Palgrave Macmillan.

Withey, S. B. (1959) 'Public opinion about science and the scientist', *Public Opinion Quarterly*, 23: 382–8.

Wynne, B. (1993) 'Public uptake of science: a case for institutional reflexivity', *Public Understanding of Science*, 2: 321–38.

—— (1995) 'Public understanding of science', in Jasanoff, S., Markle, G., Pinch, T. and Petersen, J. (eds) *Handbook of Science and Technology Studies*, London: Sage, 361–88.

Ziman, J. (1991) 'Public understanding of science', *Science, Technology and Human Values*, 16: 99–105.

—— (1992) 'Not knowing, needing to know, and wanting to know', in Lewenstein, B. (ed.) *When Science Meets the Public, Proceedings of AAAS Workshop, 17 February 1991, Washington, DC*, Washington, DC: American Association for the Advancement of Science, 13–20.

9

Scientists as public experts

Hans Peter Peters

Introduction

The function of science in society is the creation of special knowledge with a specific scientific methodology. Knowledge produced in this way is used in various forms: to build scientific theories; to develop new technologies such as medical therapies, weapons of war and nanomaterials; and to inform opinions and guide behaviour, decision-making and problem-solving. Many problems requiring the innovative use of scientific knowledge are discussed in public because they are either policy issues demanding collective decisions and the democratic involvement of citizens, or individual problems common to large groups of people. Prediction of global climate change, assessment of risks and benefits of nanotechnology, and regulation of food biotechnology are examples of such policy problems. Reduction of dietary risks and protection against behaviour-related infectious diseases such as AIDS exemplify individual problems that are common to many people.

Problems such as these are frequently addressed in public talks, exhibitions and other public events. But under the conditions of the 'media society', the mass media are particularly formative for the public sphere. Insofar as scientists are involved in the public communication of these topics – being interviewed by journalists, for example – these scientists may legitimately be called 'public experts'.

It is important to distinguish scientists as public experts from other possible roles scientists may take in public. In addition to the communication of scientific expertise – the use of scientific knowledge in the public reconstruction of non-scientific 'problems' (such as climate change) – there are two other types of science communication in which scientists are involved. The first is popularisation of research as the public reconstruction of scientific projects, discoveries, achievements and theories from a science-focused point of view; the second is meta-discourses about science and technology and the science–society relationship, such as disputes about risky technologies, and conflicts between science and social values, for example in the case of animal experimentation and research with human embryonic stem cells.

Looking at scientists as public experts combines two interesting perspectives that have been studied extensively by scholars: scientists as (policy) advisors (Jasanoff 1990; Bechmann and Hronszky 2003; Maasen and Weingart 2005) and scientists as public communicators (Friedman et al. 1986; Peters 1995; Willems 1995; Weingart 2005). Policy advice and public communication both challenge scientific norms and present dilemmas for scientists (Mohr 1996; Leshner 2007). Both tasks are interdependent and cannot be dealt with separately. Public communication of scientific expertise often has political impacts and – in response – politics, organisations and groups with political goals try to govern the production and use of scientific expertise (Stehr 2005).

Because of its practical relevance, scientific expertise is an attractive object for reporting in journalism. It is frequently covered by the media – even outside specialised science sections. In the case of global climate change, for example, studies in several countries have shown that scientific sources are prominently represented in media coverage (Wilkins 1993; Bell 1994). The same has been found for reporting on biotechnology (Bauer et al. 2001; Kohring and Matthes 2002). Food biotechnology, stem cell research, bird flu and nuclear safety are further examples of issues in which scientists are actively involved in the construction of a social reality by means of public communication. Because of its public character, that 'reality' cannot be ignored by policy-makers. Through the public sphere, scientific expertise – transformed by the logic of mass media – enters the realm of policy-making (Heinrichs et al. 2006).

The expert role of scientists in public communication

Scientific knowledge and scientific expertise

Scholars of science make a clear distinction between scientific knowledge *per se* and scientific expertise (e.g. Horlick-Jones and De Marchi 1995). Scientific knowledge *per se* is essentially concerned with the understanding of cause–effect relationships. Its concepts and theories tend to be general, abstracting as far as possible from specific situations, observations and experiments. Expert knowledge, in contrast, is concerned with the analysis and solution of practical problems in specific situations. It relates to the provision of concrete advice in specific situations to decision-makers. Successful action undoubtedly requires anticipation of the consequences of this action; to this extent, causal knowledge is a necessary component of expertise. But expertise goes beyond scientific understanding: it is concerned with the explanation of practical problems and giving advice to 'clients' who are responsible for solving these problems. It should be noted, however, that science is not the exclusive source of expertise, and that scientific knowledge, even if relevant, is usually not a sufficient source of expertise (Collins and Evans 2002). Two aspects of the basic definition of scientific expertise as advice to decision-makers based on scientific inquiries and knowledge require our attention: its relation to decision-making processes, and its provision within a social expert–client relationship (its dependence on social recognition).

Scientific knowledge and decision-making

Expertise, by definition, is provided in the context of decision-problems (including problems of orientation, forming opinions and acting). The decision-problems may be either individual problems regarding, for example, understanding a disease and choosing a medical therapy; or policy problems, such as regulation of food biotechnology or improvement of coastal protection in view of rising sea levels. Normative decision models distinguish decision options, and their expected consequences, from preferences. The models claim that a rational decision maximises the 'utility' of the outcome – they suggest selecting the option that gives the highest benefit or the lowest cost as measured against the decision-makers' preferences (Keeney and Raiffa 1976). In these models, the expert as adviser to the decision-maker has three functions (cf. Jungermann and Fischer 2005): making the client's implicit preferences explicit; developing possible decision options; and determining and assessing consequences using clients' preferences. Setting the preferences, however, remains the privilege and responsibility of the decision-maker.

Applied to politics, this normative decision-model resembles Weber's (1919) concept of the relationship between civil service and politics, according to which the professional civil service, including scientific experts, is responsible for consultation on and implementation of political decisions, and politics is responsible for setting the goals and for decision-making. Habermas (1966) has called this concept the 'decisionistic model' and distinguishes it from the 'technocratic model' – in which science *de facto* makes the decisions and politics only accepts them – and from a 'pragmatic model'. The latter model, preferred by Habermas, proposes a critical interrelationship of science and politics rather than a strict division of functions.

There is a broad consensus among scholars that the decisionistic model does not properly describe the empirical reality of science–policy interactions. However, this model seems to fit best with 'mainstream' self-descriptions of science. It is thus often implicitly referred to in the 'boundary work' of scientists (Gieryn 1983), in their attempts to defend the borders of science when reflecting about their relations with politics and the public (Mohr 1996; Leshner 2007). However, empirical analyses of scientific expertise in controversies and policy issues clearly show that the expertise delivered by scientists *de facto* is not value-free (Mazur 1985), and that scientific experts in public not only provide knowledge or comment on knowledge claims, but also evaluate options, decisions and policies, and demand strategies of political or personal action based on their own values (Nowotny 1980).

Apart from the value problem, science as expertise faces the challenge of how to deal with uncertainty. In principle, uncertainty – even controversy – is not a problem for science. Research will continue until the uncertainties are resolved and a consensus is reached. This may take quite a while, but in the end – so scientists are convinced – the research process will lead to unambiguous knowledge. In the case of expertise, however, decisions may be urgent and cannot be postponed until all uncertainty is resolved. And this pressing situation may be the rule rather than the exception.

A purist communication strategy for scientists in the face of uncertainty would be to refrain from making any factual claims in public that are not certain, thus preventing public controversies between scientists (Mohr 1996). Some authors advocate a more open approach towards uncertainty in scientific expertise. Funtowicz and Ravetz (1991), for example, distinguish several realms of science according to the degree of uncertainty involved. They develop the concept of a 'second-order science', where 'facts are uncertain, values in dispute, stakes high and decisions urgent'. They consider that a central task for experts in this field is the management of uncertainties, rather than the provision of unequivocal certainties. Böschen and Wehling (2004) argue that the ways of dealing with 'non-knowledge' are important characteristics of scientific epistemic cultures. Explicit and 'rational' ways of dealing with uncertainties are thus central elements of scientific expertise.

The media implicitly or explicitly construct the certainty or uncertainty of expert knowledge in several forms: by communicating (or omitting) explicit reservations when referring to expert knowledge; by challenging expert knowledge with non-scientific knowledge (e.g. common sense); or by quoting several expert sources that either agree or contradict each other. An important journalistic strategy for communicating scientific uncertainty in the mass media is framing it as an expert controversy.

Expert–client relationship

In psychological expert research, 'experts' are defined by their particularly high competence in a certain field. The counterpart to an expert is the novice or lay person. For sociologists, however, the expert role is defined not only by possession of special knowledge (that not everyone can be expected to have), but also by the function of giving advice (a knowledge-based service) to a client (Peters 1994). The social complement to the 'expert' is thus the 'client': somebody only becomes an expert on the basis of social recognition. In addition to competence, clients also expect loyalty from 'their' experts. And that raises the question of trust in experts – advice is usually accepted only from experts that are trusted.

In the case of scientists as public experts, the expert–client relationship is largely implied. Sometimes, scientists who are quoted in the media may intend to give advice to specific clients such as policy-makers, citizens, patients or consumers. More often, however, journalists put scientists in the expert role, relating their knowledge to policy issues or individual problems – mostly with the implicit consent of the scientists interviewed, but sometimes to their surprise.

A common problem of all forms of science communication is relating it to the relevance structure of the audience – giving the audience a good reason to 'listen' to the communication offer. The esoteric character of modern science, its incomprehensibility and detachment from everyday culture, makes it particularly difficult to connect scientific knowledge to everyday discourses and common sense. The 'mad scientist' scheme (Haynes 2003), and scientific 'miracles' and practical applications of science (Fahnestock 1986), are examples of semantic structures used by journalism to construct connections between science and the everyday world. Scientific expertise relates to the everyday world particularly well by addressing well-known and relevant problems.

The relatively easy solution of the problem of relevance makes the reporting of scientific expertise attractive for journalism. Furthermore, unlike popularisation, reporting about scientific expertise addresses not only a special science-attentive audience, but also larger audiences interested in practical problems of, for example, health, environment and technical risk. Scientific expertise is thus not confined to science sections or science programmes, but is often included in the general news coverage.

Scientific expertise and other forms of knowledge

Competition between knowledge systems

By definition, scientific expertise is scientific knowledge applied to the understanding and solution of practical problems. Sometimes these problems are known only because of science. Without science, for example, we would not know about the ozone hole or global climate change. More often, however, the problems are obvious or already well-known by experience, and more-or-less successful strategies to cope with them are in place. When scientists offer their expertise in a 'pluralistic knowledge society' (Heinrichs 2005), their knowledge frequently meets competing knowledge from other research communities and extra-scientific domains: everyday knowledge, special knowledge based on practical experience, or traditional knowledge stemming, for example, from religion, folk wisdom or indigenous culture.

Scientific knowledge's competition with other knowledge forms causes two kinds of problem. First, prior knowledge about a subject may hinder the understanding and acceptance of new knowledge, leading to problems of explanation. Rowan (1999) has analysed such cognitive problems in lay people's acquisition of scientific knowledge. As an explanatory strategy to overcome that problem, Rowan recommends making explicit the contradiction between everyday knowledge and scientific knowledge. Second, the question arises as to which of the competing knowledge forms – for example, scientific knowledge and everyday knowledge – is more valid and more suitable for solving the problem. Wynne (1996), along with many others, has challenged the assumption that scientific knowledge is *per se* superior to 'local knowledge' based on experience.

If problem-relevant everyday knowledge exists in competition with scientific knowledge, scientists face the risk that the media either ignore them, or occasionally characterise them as theorists who are removed from the real world and without a proper understanding of the practical problem. In this context, the media may often point to a specific case in which scientific advice has failed to solve the problem. Or, if common sense and scientific expertise are congruent, the media may sometimes reproach scientists for spending much time (and public money) only to find the obvious.

Integration of knowledge – towards a broader concept of scientific expertise?

It is obvious that the solution of practical problems requires several kinds of knowledge. Even if valid and relevant scientific theories are available, the conditions and constraints of the specific case have to be known in order to 'calculate' the effects of interventions. Furthermore, scientific theories usually deal with idealised (simplified) models of, for example, the human body, the global climate or the system of agriculture. The concrete 'real systems' may differ from these idealised models in several respects. Detailed knowledge of the characteristics of the concrete system is required. In many cases, systems are too complex to be understood completely. From long-term experience with systems, experts often nevertheless possess problem-relevant 'implicit knowledge' (Perrig 1990).

That scientific knowledge usually is insufficient as the sole basis of expertise becomes even more obvious when we consider the development of decision options and strategies for action. Intimate knowledge about available resources, legal, political and psychological constraints and implementation barriers, for example, is necessary to devise efficient action strategies. A climate researcher recommending in a media interview that energy consumption be reduced by half in the next 10 years will immediately provoke the journalist's next question: 'How?'

In complex problems, we cannot expect a single expert to be competent in all aspects. The need to use different sources of knowledge and to integrate them in order to solve a practical problem is obvious. The demand for integrated expertise has consequences for knowledge management, including journalistic strategies of inquiry, but also for knowledge production. In the current debate about research strategies to deal with ecological problems, for example, there is an almost ubiquitous call for more 'interdisciplinary' or 'transdisciplinary' research that exceeds the boundaries of scientific disciplines or even the classical framework of science at large (Somerville and Rapport 2003). Gibbons et al. (1994) claim the emergence of a new mode of knowledge production (mode 2), which involves a broader spectrum of actors such as practitioners and users of knowledge, takes place outside academic institutions, and leads to more contextualised and 'socially robust' knowledge (Nowotny et al. 2001). Grunwald (2003) even discusses the possibility of 'transsubjective' normative advice as part of scientific expertise.

We may view the described trends in (transformed) science as expanding beyond the classical domain of science in two ways: first, the change in the character of science may be understood as the attempt to usurp total responsibility for problem-definition and problem-solution as a kind of 'second-order technocracy'; second, the change may be seen as allowing contextualised knowledge, social values and interests to enter the scientific knowledge-generation process in a controlled way, and at an early stage, thus decreasing the need for *ex post* reintegration of non-scientific knowledge, values and interests in the development of expertise.

Contextualising scientific knowledge, confronting it and complementing it with non-scientific forms of knowledge may be an important task of (science) journalism. Discussing two different forms of rationality and their respective advantages and disadvantages for problem-solving, Spinner (1988) calls journalists 'agents of occasional reason' and assigns them the task of challenging the principle-based scientific–technical rationality with 'occasional rationality': with the cunning orientation at specific, local and temporary aspects of events and situations.

The science–journalism interface

Selection of expert sources

Several researchers have looked at the selection of scientific experts by the media. In a case study of reporting of the marijuana controversy, Shepherd (1981) found that the media quoted as experts not primarily the most relevant and experienced researchers, but rather health administrators and highly prominent scientists, regardless of their specific field of expertise. From the perspective of journalists, it is not research productivity but other qualities that define a good public expert. Practitioners as well as senior scientists with overview knowledge and general experience may be better suited than the actual researchers in the subject matter to relate research to decision problems, to integrate different knowledge sources, and to provide contextualised expertise. Rothman (1990) has analysed possible biases in the selection of experts from a scientific community. He concludes from several case studies of expert controversies that journalists' selection of experts is biased: experts representing minority positions are usually overrepresented in the coverage. Kepplinger et al. (1991) argue that media tend to select expert sources that support their editorial policies. Goodell (1977: 4) concludes that the media focus on relatively few 'visible scientists' and select scientific sources 'not for discoveries, for popularising, or for leading the scientific community, but for activities in the tumultuous world of politics and controversy'.

Selecting expert sources for journalists' purposes is a complex process in which scientific productivity and reputation are not the only factors. The main journalistic criterion in the selection of sources is whether a source makes a good story or improves a story. What makes a good story, however, differs between media, between sections and programmes, and between different topics. Some of the more important factors that influence the likelihood of scientists appearing in the media are listed below.

- Relevance: a scientific source must be able to comment on something relevant for the audience. In some cases the relevance is quite obvious, as, for example, with medical therapies, environmental risks or government decisions. However, this criterion is not as restrictive as it may seem, as journalists can construct 'relevance' in several ways. On a very general level, the 'news values' concept (Fuller 1996) describes some of the criteria journalists use to assess public relevance.
- Visibility: scientists become visible to journalists by their involvement in events and debates outside science (e.g. as members of policy advisory boards or authors of expert opinions commissioned by the government). They are more visible if they publish in journals (such as *Science* and *Nature*) or talk at conferences (such as the Annual Meeting of the American Association for the Advancement of Science) that are regularly monitored by journalists. Furthermore, the public relations of scientific organisations, journals, associations and congresses strongly influence the visibility of scientists (Peters and Heinrichs 2005). Finally, prior media coverage makes scientists more visible. This leads to the feedback loops of media attention resulting in the 'visible scientists' described by Goodell (1977).
- Accessibility and media appropriateness: journalists work with limited resources and in a narrow time frame. The anticipated effort required to deal with a scientist is an important selection criterion. Whether scientists reply to emails and messages, how quickly they reply, and how 'complicated'

they are in the interactions therefore matters. Journalists prefer scientists who are able and willing to speak crisply and concisely, to answer the questions asked, to explain complicated matters using comparisons and metaphors, and to draw bold conclusions. Furthermore, journalists (usually) prefer scientists with high organisational rank and public reputation.

Interactions of scientific experts and journalists

There are several reasons to expect that the relationship between scientists and journalists is tense and the interactions difficult. Intercultural communication theory predicts misunderstandings caused by the cultural difference between science and journalism (Peters 1995). Game theory lets us expect interest conflicts between scientists and journalists because of partly incongruent goals and competing strategies. From the social systems theory point of view, scientific meaning is confined to the science system, thus precluding a simple 'transfer' or 'translation' of messages from science to the public. Journalistic constructs are the result of journalistic 'observation'. Because of the different 'logics' of science and public communication, scientific and journalistic constructs of the same research necessarily differ (Kohring 2005).

A number of surveys have studied scientists' attitudes towards the media and their contacts with journalists (Dunwoody and Ryan 1985; Hansen and Dickinson 1992; Kyvik 1994; Willems 1995). The following account of the interactions of scientists and journalists is based mainly on two German studies. These explicitly address scientists as public experts and consist of matching surveys of scientists and journalists who actually had contact in the preparation of media reports (Peters 1995; Peters and Heinrichs 2005). The first of these studies focused on risk reporting in general; the second study looked at the interactions of experts and journalists in reporting global climate change. With some consistency, the two studies found several systematic differences in the mutual expectations of scientists and journalists (cf. Peters 2007).

- Communication norms: scientists tend to apply scientific communication norms to public communication. They prefer to focus on knowledge in their specialist field and – compared with journalists – they like a serious, matter-of-fact, cautious and educational style of communication. Their journalistic interaction partners do not completely disagree, but look for overview knowledge, preferring clear messages, evaluative comments and an entertaining style.
- Model of journalism: scientists favour a kind of 'service model', normatively expecting journalists to help them promote scientific goals and interests. Based on their professional norms, journalists, at least verbally, insist on keeping distance from the objects they are reporting, on their independence, and on a watchdog perspective.
- Control of communication: a clear-cut disagreement exists in the issue of control: who should control the communication with the public and the media content? Journalists consider themselves as responsible authors, and

> consider the scientist as their 'source' – as a resource for their task of writing a story. According to journalism norms, journalists owe sources fair treatment (e.g. correct quotation) but nothing more. In particular, they are very critical of demands from sources that may be viewed as censorship. Scientists, however, think that they are the real authors and should control the communication process because they are the originators of the message to be conveyed to the public. In accordance with their service model of journalism, they tend to assign journalists a role as disseminator only.

Despite these disagreements, the general conclusion from the two German surveys points to a surprisingly strong co-orientation of scientific experts and journalists. In many respects, the expectations of both groups are actually congruent: in some respects there are only moderate discrepancies, and only regarding the control of communication is there outright disagreement. Scientists and journalists seem to possess effective strategies to overcome the problems arising from the discrepant expectations. Even more, the surveys show that scientists, to a large degree, not only accept the journalistic treatment of science but anticipate it in their statements. In both surveys, journalists and scientists express a high degree of satisfaction with their mutual interactions, and the experts – although noticing many inaccuracies and changes of meaning – are, to a large extent, satisfied with the media stories in which they are quoted.

Public (re-)construction of scientific expertise

Public communication of science cannot be understood as 'translation'. Translation would require structural equivalence of source and target language, and a shared reality serving as background for making sense of information. There is neither equivalence of scientific and everyday language, nor a shared reality. The worlds of modern science are esoteric and rather inaccessible to everyday reasoning.

Scientists and the media therefore have to construct public images of the esoteric worlds of science and their events using terms, metaphors, comparisons and concepts from the everyday world. With respect to public expertise, this means that while expert *advice* itself can be expressed in everyday language and refers to the objects and events of an everyday world, the scientific justification of that advice will often be incomprehensible. The validity of scientific expert advice therefore cannot be proven to clients, but only made plausible. In the end, clients must trust their experts. To inform trust, however, the social context of expertise (e.g. interests, neutrality, independence) becomes crucial, and a legitimate field of journalism inquiry and reporting (Kohring 2004).

The analysis of how scientific expertise in the media is constructed is an urgent research field. Some general features of public constructs of science and scientific expertise can be mentioned, however. The neglect of scientific detail and accuracy, demonstrated by numerous studies (e.g. Singer 1990, Bell 1994), is not an indicator of unprofessional journalism but, on the contrary, the consequence of journalistic professionalism: journalists do not adopt the quality criteria of science, such as accuracy, but follow their own criteria (Salomone et al. 1990). Scientific errors and inaccuracies are only the 'tip of the iceberg' of semantic discrepancies between scientific and public constructs of science.

Dunwoody (1992) and others use the concept of 'story frames' to explain which kind of information journalists include or exclude from their stories. These frames guide the organisation of information in the story, but also journalistic inquiry. Two-thirds of the journalists interviewed in the survey of climate change experts and journalists by Peters and Heinrichs (2005) said that they had the outline of the story in mind when contacting a scientist, and one-third of the scientists said they had the impression that the interviewing journalist wanted to hear something particular. Journalists often have pronounced expectations, not only with respect to the topic they ask their sources to comment about, but also regarding the stances their sources take.

Comparing popular science stories written by scientists and by journalists about the same research, Fahnestock (1986) found several systematic differences that illustrate some of the journalistic rules of constructing public science. She found, for example, that journalists focused more than scientists on explaining the purpose of research, and that pointing to practical applications was one possible answer. She also found that journalists did not focus as much as scientists did on the scientific results of research, but rather looked at its consequences for evaluation and action. Peters and Heinrichs (2005) analysed journalists' use of information provided by climate-change experts in media interviews, and concluded that journalists expect problem-oriented, interpreted knowledge rather than pure scientific research results. Journalists contextualised information received from scientists by relating it to concrete events or problems. Furthermore, as also observed by Fahnestock (1986), journalists omitted reservations and thus let conclusions appear less uncertain and more general than the scientists had intended. In many cases, journalism not only makes scientific expertise public and emphasises the expertise-character of research, but actively contributes to the creation of 'public expertise'. Journalists confront their scientific sources with specific events (e.g. weather anomalies), asking them to explain these, and they urge them to comment on policy problems.

Legitimatory and persuasive uses of public expertise

Scientific authority and trust

Turner (2001: 124) argues that the state accepts the authority of science and expertise by 'requiring that regulations be based on the findings of science or on scientific consensus'. Moreover, public opinion surveys in Europe and the USA show that scientific institutions are among the most trusted societal institutions, and that, compared with other professions, scientists possess a high credibility (National Science Board 2004; European Commission 2005). In an analysis of climate-change media coverage, Peters and Heinrichs (2005) found that, on average, scientific sources were rated more positively than non-scientific sources.

Several authors nevertheless diagnose a credibility crisis of scientific expertise (e.g. Horlick-Jones and De Marchi 1995). The high general trust in scientific expertise is challenged in contexts in which scientists are perceived as interest group or as advocates for a technology (Peters 1999), or in which experts publicly disagree (Rothman 1990). Media reception studies show that recipients of media stories generate critical thoughts about experts quoted in these stories if the experts express opinions that contradict the pre-existing attitudes of the recipients (Peters 2000). Trust is not unconditionally assigned to science and scientists, but rather is modified and specified dependent on the context, taking into account such things as organisational affiliation (e.g. academia versus industry) and advocacy relationships of scientists.

Giddens (1991) emphasises the ubiquity of trust in 'expert systems' and in the experts that operate these. Although there are several fields of political controversy related to controversial expertise, the usual way of dealing with routine problems of a technical matter is the application of scientific expertise in a non-political (technocratic) way. Trust in science and scientific expertise is still the default, unless – in few selected domains – its authority is purposefully dismantled by organised distrust and counter-expertise – both recommended by Beck (1988) as strategies against 'risk technocracy'. The authority of science is so strong that it usually can be neutralised only by counter-expertise. Critics of established scientific expertise have to refer to science in order to effectively support their claims. In the USA, for example, the opponents of Darwin's theory of evolution formulate their largely religiously motivated alternative as a quasi-scientific theory ('intelligent design') in order to increase its persuasiveness (Park 2001). Scientific controversies are thus proof more of the strong legitimatory and persuasive effects of science than of its declining authority. Promoted by the professionalisation of the environmental movement, however, the use of science to legitimate certain points of view and to persuade is no longer monopolised by the state and industry. This is sometimes referred to as 'democratisation of expertise' (cf. Maasen and Weingart 2005).

Advocacy science and expert controversies

Many public controversies refer to science and technology and scientists participate as advocates on both sides of the controversy (Mazur 1981; Frankena 1992; Nelkin 1992). In such controversies, science can be involved in different ways: scientific research or science-based technologies, e.g. nuclear power or genetic engineering, may be the controversial issue or the definition of problem, and the assessment of pros and cons of policy options may depend on scientific expertise. In both cases, control over both the definition ('who is a legitimate expert?') and the content ('which experts are prepared to "testify" in the media?') of public expertise is crucial for the opening and closure of controversies. Nowotny (1980) and others have found that the readiness of scientists to serve as public experts depends partly on their position in the controversy. Experts challenging the establishment are usually more prepared to participate in public debates than those defending a technology or a government policy. Experts are therefore an important power resource for non-governmental organisations (NGOs). As legitimate journalistic sources, they help secure access to the public sphere, they increase the rationality of the claims, they provide legitimacy drawn from scientific authority, and they contribute to the construction of a social reality that favours the acceptance of their clients' claims. In order to have better access to, and control over, the content of expertise, NGOs have developed an infrastructure of research institutes and of networks of sympathising experts providing 'critical expertise' or 'counter-expertise'.

In controversies, scientific expertise can become a persuasive resource, a means to convince the population and political decision-makers of a certain 'reality' and to act in a specific way (Heinrichs et al. 2006). Once a social reality is defined (e.g. 'global change is a relevant risk'), the legitimacy of actors, claims and policies is judged against that reality. NGOs, for example, may refer to the scientific description of climate change and request changes in energy policy. The science used in these processes, in which scientific authority is transformed into political legitimacy, is labelled 'advocacy science' – the coalition of science with political, economic or other interests. The 'legitimating role of expert advice' (Rip 1985), that is, the use of scientific evidence to rationalise and justify rather than to make decisions, has repercussions for the public image of science and even, perhaps, for scientific communities and processes (Weingart 2005).

Expertise as part of science PR

Scientists are members of organisations, for example, universities or government research laboratories, and their situation in those organisations affects the communication behaviour of individual researchers (Dunwoody and Ryan 1983) and thus influences the communication of public expertise. Many organisations have informal or formal guidelines that regulate contacts between individual scientists and the media, and use their public relations efforts to promote their goals.

Research organisations have to legitimise their use of societal resources (e.g. money and manpower), their demand of autonomy and, in some cases, also the conflict between their research and public values (e.g. in the case of animal experiments and use of embryonic stem cells). There are several principal ways in which scientific organisations can create such a legitimacy by their public relations:

- appealing to legal or moral commitment, they can argue that they are entitled to societal support;
- demonstrating status, success and conformity with social values, they can show that they deserve support; and
- pointing to benefits of their work for society, they can suggest that it is in the self-interest of society to provide support.

Improving the public debate of policy issues and common individual problems through scientific expertise, and encouraging their scientists to take on the role of public experts, is usually in the interest of research organisations. In a media society, public visibility is generally an indicator of relevance and therefore of 'status' and 'success'. Provision of expertise that helps to understand and solve policy issues and individual problems is an important way of increasing the utility of research for society and thus demonstrating the beneficial impact of the respective organisations. With few exceptions, research organisations therefore try to increase their visibility for the policy-makers, funding bodies and tax payers. If possible, organisational PR focuses on research with practical applications and on scientific expertise, for two reasons: because it is particularly easy to connect it to political and everyday discourse, and because the benefits of research for society are most obvious in these areas. Many public relations departments keep lists of experts who are ready to talk to the media. And some organisations even 'advertise' on their websites their scientists who might comment on current public issues, as a service for journalists who are looking for suitable interview partners.

However, the media sometimes use scientific experts to challenge the political establishment, criticising its policy, for example. In such a case, a scientific organisation may suddenly find itself in conflict with the government – not desirable for a university president or the director of a research centre. While, in general, organisations encourage scientists to act as public experts, they are probably more reserved if they expect the expertise to challenge governmental (or other funding bodies') policies and interests.

Concluding remarks

One of the roles scientists take, or are urged to take, in public, in particular in the mass media, is that of expert. Compared with pure scientific knowledge, expertise is defined by its reference to social problems, decision-making and action. Being framed as public experts implies the expectation that scientists apply their knowledge to the explanation and solution of non-scientific problems.

On the one hand, providing expertise is rewarding for scientists because, in contrast to esoteric scientific discoveries or theories, expertise usually connects rather easily to what the media and their audiences consider relevant. On the other hand, being an expert means crossing the boundary of science, entering society as an actor, and exposing oneself to internal and external criticism. As experts, scientists do not possess a monopoly of relevant knowledge; values and interests will come into play and public controversies may evolve.

While some scientists are prepared to become involved in issues of policy or public health, other scientists may be reluctant to enter the 'tumultuous world of politics and controversy' (Goodell 1977: 4). Journalists tend to focus on the connections of scientific knowledge with the non-scientific world, and will often push scientists in interviews to the limit of (or even beyond) what they are prepared to offer about the practical implications of their knowledge. Journalism thus has an important function not only in the public communication of scientific expertise, but also in its creation.

Selected further reading

Horlick-Jones, T. and De Marchi, B. (1995) 'The crisis of scientific expertise in *fin de siècle* Europe', *Science and Public Policy*, 22: 139–45.

Maasen, S. and Weingart, P. (eds) (2005) *Democratization of Expertise? Exploring Novel Forms of Scientific Advice in Political Decision-Making*, Dordrecht: Springer.

Peters, H. P. (1995) 'The interaction of journalists and scientific experts: co-operation and conflict between two professional cultures', *Media, Culture & Society*, 17: 31–48.

Rip, A. (1985) 'Experts in public arenas', in Otway, H. and Peltu, M. (eds) *Regulating Industrial Risks*, London: Butterworth, 94–110.

Spinner, H. F. (1988) 'Wissensorientierter Journalismus. Der Journalist als Agent der Gelegenheitsvernunft', in Erbring, L. (eds) *Medien ohne Moral?*, Berlin: Argon, 238–66.

Weingart, P. (2005) *Die Wissenschaft der Öffentlichkeit: Essays zum Verhältnis von Wissenschaft, Medien und Öffentlichkeit*, Weilerswist: Velbrück.

Further references

Bauer, M., Kohring, M., Allansdottir, A. and Gutteling, J. (2001) 'The dramatisation of biotechnology in elite mass media', in Gaskell, G. and Bauer, M. W. (eds) *Biotechnology 1996–2000 – the Years of Controversy*, London: Science Museum, 35–52.

Bechmann, G. and Hronszky, I. (eds) (2003) *Expertise and its Interfaces. The Tense Relationship of Science and Politics*, Berlin: edition sigma.

Beck, U. (1988) *Gegengifte. Die organisierte Unverantwortlichkeit*, Frankfurt: Suhrkamp.

Bell, A. (1994) 'Media (mis)communication on the science of climate change', *Public Understanding of Science*, 3: 259–75.

Böschen, S. and Wehling, P. (2004) *Wissenschaft zwischen Folgenverantwortung und Nichtwissen*, Wiesbaden: Verlag für Sozialwissenschaften.

Collins, H. M. and Evans, R. (2002) 'The third wave of science studies: studies of expertise and experience', *Social Studies of Science*, 32: 235–96.

Dunwoody, S. (1992) 'The media and public perceptions of risk: how journalists frame risk stories', in Bromley, D. W. and Segerson, K. (eds) *The Social Response to Environmental Risk*, Boston. MA: Kluwer, 75–100.

Dunwoody, S. and Ryan, M. (1983) 'Public information persons as mediators between scientists and science writers', *Journalism Quarterly*, 66: 647–56.

—— (1985) 'Scientific barriers to the popularization of science in the mass media', *Journal of Communication*, 35: 26–42.

European Commission (2005) *Social Values, Science and Technology. Eurobarometer 2005*, Brussels: European Commission.

Fahnestock, J. (1986) 'Accommodating science: the rhetorical life of scientific facts', *Written Communication*, 3: 275–96.

Frankena, F. (1992) *Strategies of Expertise in Technical Controversies: A Study of Wood Energy Development*, Bethlehem, PA: Lehigh University Press.

Friedman, S. M., Dunwoody, S. and Rogers, C. L. (eds) (1986) *Scientists and Journalists. Reporting Science as News*, New York and London: The Free Press & Macmillan.

Fuller, J. (1996) *News Values: Ideas for an Information Age*, Chicago, IL: University of Chicago Press.

Funtowicz, S. O. and Ravetz, J. R. (1991) 'A new scientific methodology for global environmental issues', in Costanza, R. (ed.) *Ecological Economics: The Science and Management of Sustainability*, New York: Columbia University Press, 137–52.

Gibbons, M., Limoges, C., Nowotny, H., Schwartzman, S., Scott, P. and Trow, M. (1994) *The New Production of Knowledge*, London: Sage.

Giddens, A. (1991) *The Consequences of Modernity*, Cambridge: Polity Press.

Gieryn, T. F. (1983) 'Boundary work and the demarcation of science from non-science: strains and interests in professional ideologies of scientists', *American Sociological Review*, 48: 781–95.

Goodell, R. (1977) *The Visible Scientists*, Boston, MA: Little, Brown & Co.

Grunwald, A. (2003) 'Methodological reconstruction of ethical advices', in Bechmann, G. and Hronszky, I. (eds) *Expertise and its Interfaces. The Tense Relationship of Science and Politics*, Berlin: edition sigma, 103–24.

Habermas, J. (1966) 'Verwissenschaftlichte Politik in demokratischer Gesellschaft', in Krauch, H., Kunz, W. and Rittel, H. (eds) *Forschungsplanung*, München: Oldenbourg, 130–44.

Hansen, A. and Dickinson, R. (1992) 'Science coverage in the British mass media: media output and source input', *Communications*, 17: 365–77.

Haynes, R. (2003) 'From alchemy to artificial intelligence: stereotypes of the scientist in Western literature', *Public Understanding of Science*, 12: 243–53.

Heinrichs, H. (2005) 'Advisory systems in pluralistic knowledge societies: a criteria-based typology to assess and optimize environmental policy advice', in Maasen, S. and Weingart, P. (eds) *Democratization of Expertise?*, Dordrecht: Springer, 41–62.

Heinrichs, H., Petersen, I. and Peters, H. P. (2006) 'Medien, Expertise und politische Entscheidung: das Fallbeispiel Stammzellforschung', in Wink, R. (ed.) *Deutsche Stammzellpolitik im Zeitalter der Transnationalisierung*, Baden-Baden: Nomos, 119–40.

Horlick-Jones, T. and De Marchi, B. (1995) 'The crisis of scientific expertise in *fin de siècle* Europe', *Science and Public Policy*, 22: 139–45.

Jasanoff, S. (1990) *The Fifth Branch: Science Advisors as Policymakers*, Cambridge, MA: Harvard University Press.

Jungermann, H. and Fischer, K. (2005) 'Using expertise and experience for giving and taking advice', in Betsch, T. and Haberstroh, S. (eds) *The Routines of Decision Making*, Mahwah, NJ: Lawrence Erlbaum, 157–73.

Keeney, R. and Raiffa, H. (1976) *Decisions with Multiple Objectives: Preferences and Value-tradeoffs*, New York: Wiley.

Kepplinger, H. M., Brosius, H.-B. and Staab, J. F. (1991) 'Instrumental actualization: a theory of mediated conflicts', *European Journal of Communication*, 6: 263–90.

Kohring, M. (2004) *Vertrauen in Journalismus. Theorie und Empirie*, Konstanz: UVK.

—— (2005) *Wissenschaftsjournalismus. Forschungsüberblick und Theorieentwurf*, Konstanz: UVK.

Kohring, M. and Matthes, J. (2002) 'The face(t)s of biotech in the nineties: how the German press framed modern biotechnology', *Public Understanding of Science*, 11: 143–54.

Kyvik, S. (1994) 'Popular science publishing', *Scientometrics*, 31: 143–53.

Leshner, A. I. (2007) 'The evolving context for science and society', in Claessens, M. (ed.) *Communicating European Research 2005*, Dordrecht: Springer, 27–32.

Maasen, S. and Weingart, P. (eds) (2005) *Democratization of Expertise? Exploring Novel Forms of Scientific Advice in Political Decision-Making*, Dordrecht: Springer.

Mazur, A. (1981) *The Dynamics of Technical Controversy*, Washington, DC: Communications Press.

—— (1985) 'Bias in risk–benefit analysis', *Technology in Society*, 7: 25–30.

Mohr, H. (1996) 'Das Expertendilemma', in Nennen, H.-U. and Garbe, D. (eds) *Das Expertendilemma. Zur Rolle wissenschaftlicher Gutachter in der öffentlichen Meinungsbildung*, Berlin: Springer, 3–24.

National Science Board (2004) *Science and Engineering Indicators 2004*, Arlington, VA. USA: National Science Foundation.

Nelkin, D. (ed.) (1992) *Controversy. Politics of Technical Decisions*, 3rd edn, Newbury Park, CA: Sage.

Nowotny, H. (1980) 'Experten in einem Partizipationsversuch. Die Österreichische Kernenergiedebatte', *Soziale Welt*, 31: 442–58.

Nowotny, H., Scott, P. and Gibbons, M. (2001) *Re-Thinking Science. Knowledge and the Public in an Age of Uncertainty*, Cambridge: Polity Press.

Park, H.-J. (2001) 'The creation–evolution debate: carving creationism in the public mind', *Public Understanding of Science*, 10: 173–86.

Perrig, W. J. (1990) 'Implicit knowledge – a challenge for cognitive psychology', *Schweizerische Zeitschrift für Psychologie*, 49: 234–49.

Peters, H. P. (1994) 'Wissenschaftliche Experten in der öffentlichen Kommunikation über Technik, Umwelt und Risiken', in Neidhardt, F. (ed.) *Öffentlichkeit, öffentliche Meinung, soziale Bewegungen*, Opladen: Westdeutscher Verlag, 163–90.

—— (1995) 'The interaction of journalists and scientific experts: co-operation and conflict between two professional cultures', *Media, Culture & Society*, 17: 31–48.

—— (1999) 'Das Bedürfnis nach Kontrolle der Gentechnik und das Vertrauen in wissenschaftliche Experten', in Hampel, J. and Renn, O. (eds) *Gentechnik in der Öffentlichkeit*, Frankfurt: Campus, 225–45.

—— (2000) 'The committed are hard to persuade. Recipients' thoughts during exposure to newspaper and TV stories on genetic engineering and their effect on attitudes', *New Genetics & Society*, 19: 367–83.

—— (2007) 'The science–media interface: interactions of scientists and journalists', in Claessens, M. (ed.) *Communicating European Research 2005*, Dordrecht: Springer, 53–8.

Peters, H. P. and Heinrichs, H. (2005) *Öffentliche Kommunikation über Klimawandel und Sturmflutrisiken. Bedeutungskonstruktion durch Experten, Journalisten und Bürger*, Jülich: Forschungszentrum.

Rip, A. (1985) 'Experts in public arenas', in Otway, H. and Peltu, M. (eds) *Regulating Industrial Risks*, London: Butterworth, 94–110.

Rothman, S. (1990) 'Journalists, broadcasters, scientific experts and public opinion', *Minerva*, 28: 117–33.

Rowan, K. E. (1999) 'Effective explanation of uncertain and complex science', in Friedman, S. M., Dunwoody, S. and Rogers, C. L. (eds) *Communicating Uncertainty*, Mahwah, NJ: Lawrence Erlbaum, 201–23.

Salomone, K. L., Greenberg, M. R., Sandman, P. M. and Sachsman, D. B. (1990) 'A question of quality. How journalists and news sources evaluate coverage of environmental risk', *Journal of Communication*, 40: 117–33.

Shepherd, R. G. (1981) 'Selectivity of sources: reporting the marijuana controversy', *Journal of Communication*, 31: 129–37.

Singer, E. (1990) 'A question of accuracy. How journalists and scientists report research on hazards', *Journal of Communication*, 40: 102–16.

Somerville, M. A. and Rapport, D. J. (2003) *Transdisciplinarity: Recreating Integrated Knowledge*, Oxford: EOLSS.

Spinner, H. F. (1988) 'Wissensorientierter Journalismus. Der Journalist als Agent der Gelegenheitsvernunft', in Erbring, L. (eds) *Medien ohne Moral?*, Berlin: Argon, 238–66.

Stehr, N. (2005) *Knowledge Politics: Governing the Consequences of Science and Technology*, Boulder, CO: Paradigm.

Turner, S. (2001) 'What is the problem with experts?', *Social Studies of Science*, 31: 123–49.

Weber, M. (1919) *Politik als Beruf*, München: Duncker & Humblot.

Weingart, P. (2005) *Die Wissenschaft der Öffentlichkeit: Essays zum Verhältnis von Wissenschaft, Medien und Öffentlichkeit*, Weilerswist: Velbrück.

Wilkins, L. (1993) 'Between facts and values: print media coverage of the greenhouse effect, 1987–1990', *Public Understanding of Science*, 2: 71–84.

Willems, J. (1995): 'The biologist as a source of information for the press', *Bulletin of Science, Technology & Society*, 15: 21–46.

Wynne, B. (1996) 'May the sheep safely graze? A reflexive view of the expert-lay knowledge divide', in Lash, S., Szerszynski, B. and Wynne, B. (eds) *Risk, Environment and Modernity. Towards a New Ecology*, London: Sage, 44–83.

10

Public relations in science

Managing the trust portfolio

Rick E. Borchelt

Introduction

Public relations (PR) has come to play an important and inextricable part in science communication – but one that has escaped much of the dedicated research effort that has helped scholars understand other variables, such as the role of the media, public attitudes toward science, and the impact of the internet on acquiring scientific knowledge. This chapter discusses some of the various approaches to PR that scientific organisations employ; the goals these organisations have for PR; the levels in scientific organisations at which PR is or can be employed; management of PR in terms of organisational behaviour; and ways to evaluate PR practice as both a process and an outcome. The approach taken is to view PR in a scientific organisation as 'managing the trust portfolio' – both for the organisation and for the scientific enterprise more generally, and as a unifying concept for future scholarship.

Throughout this discussion, the terms 'public relations' and 'public affairs' are used interchangeably, both referring to the communication management function of an organisation. Public relations is the term most often used by academic researchers in communications and in the corporate world, but in many situations in the scientific world, PR has come to denote a less-than-savoury bag of tricks to confuse or dupe potential customers or citizens. As I use it, PR is the art and science of developing meaningful relationships with the public necessary for continuing the work of an organisation. It is not intended as a synonym for marketing, although marketing may be a component of PR practice in some scientific organisations.

Whether they acknowledge it or not, scientific organisations use PR in a variety of ways. Research-performing organisations, such as colleges and universities, advertise the quality of their programmes to recruit new students and faculty, to publicise work being funded by government in hopes of receiving additional funding, and to share with taxpayers the results of research conducted with government support. Non-profit advocacy organisations publicise their work to attract new donors, or call attention to their issues and how they have been able to effect legislative or policy change.

Corporate scientific institutions attract and retain customers or investors, or seek to change consumer behaviour toward their or a similar product.

Yet with few exceptions – notably the singular role of the science information officer as an actor or gatekeeper in communication (Friedman et al. 1986; Nelkin 1995) – PR in science has been poorly studied by communication scholars.

Public relations and science: a brief historical outlook

Public relations as we know it today is a relatively young field, practiced as a discipline in the corporate world for only about 100 years. The world of science is a latecomer to PR, and many scientists remain unconvinced of the value of cultivating relationships or communicating with publics beyond their scientific peers. In some form, marketing and publicity have been with us throughout history. The development of effective means of mass communication in the 19th century, though, created an entirely new field of play for publicists, and this initial 'publicity' phase of PR continues to this day, even benefiting from new technologies to reach mass markets more quickly, with more targeted messages. Historians of PR also refer to this kind of PR as 'press agentry', so named because its practitioners often went by the name 'press agents' and were the marketers that hawked press releases and news tips to a willing media enterprise (Grunig and Hunt 1984).

At its heart, press agentry PR seeks to maximise awareness of a product, an idea or an institution. 'Making the news' or 'getting ink' are the primary benchmarks of success for press agentry, and 'placement' of stories about one's organisation in media outlets still is a goal desired by many scientific organisations. The number of press releases produced, how many were used by reporters, and how many stories resulted from the press releases are typical measures of success. Today, PR professionals in scientific organisations are just as likely to be asked 'how many web hits did we get on that story?' or 'how much airtime did that get on the nightly news?'

One of the early practitioners of publicity management, the circus and entertainment mogul P. T. Barnum, is credited with saying, 'I don't care what the newspapers say about me as long as they spell my name right' (cit. by Bernard and Duffy 1982). It is often an underlying assumption in press agentry that quantity trumps quality in media coverage, and a stack of newspaper clippings is often the only deliverable that PR professionals in science are asked to produce. Even negative publicity is sometimes seen as good publicity, as long as it keeps the name of the organisation and its researchers in the public eye.

What is important about press agentry is the direction of information flow: it is almost entirely one-way, from the organisation to its public or publics, and there is hardly any feedback loop from the public to the boardroom or the laboratory bench – with the exception of profits, recruitment or financial support (Grunig and Hunt 1984).

Public relations practitioners soon learned, however – even if their CEOs didn't – that public attention was no guarantee of public support. More enlightened organisations began to realise that the wrong kind of publicity had bad consequences for attracting and retaining customers, and that being known for the 'right' things was as important as simply being known. This required a certain amount of explaining to the public just what you were up to, and a new phase of PR – the 'explanatory' phase – developed early in the 20th century (Grunig and Hunt 1984).

One of the first advocates of explanatory PR was Ivy Lee, who started his career as a newspaper reporter and later came to advise John D. Rockefeller, the American Red Cross and the Pennsylvania Railroad, among others. His approach to PR was to open up the process of communication. In one famous episode, a railroad client stood accused of safety violations that resulted in a number of accidents, and was prepared to do what all big institutions of its day did: stonewall the press, keep anyone from talking, cover up the information as much as possible, and wait for the problem to blow over. The railroad had just hired Lee, who convinced them to try a different approach: bring in the reporters, explain what had happened, feed them information they could use, set them up with interviews with good spokespeople, and try to control unfavorable newspaper coverage rather than ignore it or try and make it go away (Marchand 1998).

Lee's form of explanatory public relations soon became the major form of PR practised among companies, political campaigns, government agencies, universities and non-profit organisations. In the scientific world, it is best known as public information practice, named after the title that many universities and federal agencies give its practitioners: public information officers (PIOs). While not all PIOs in science today practice simple explanatory PR, many still produce only the news releases, fact sheets, tip sheets and other materials that cast the most favourable light on their institutions' scientific endeavours. The word that many people often use in referring to explanatory PR of this type is 'spin control' – making sure the public knows a lot about the science or the scientists, but only the 'right' things that the organisation thinks the public should know.

An example of explanatory PR practice would be a government health agency putting out only information about successes in health promotion programmes, even while it is tracking or monitoring potentially devastating public health problems. In Britain, the Department of Health was accused of just such a PR approach by systematically downplaying and minimising the risk of bovine spongiform encephalopathy (BSE), or mad cow disease, when the epidemic first came to public attention (see Chapter 14 in this volume). Only after repeated media accounts of increasing numbers of victims, and the destruction of thousands of potentially infected animals, did the government finally confess that the problem was worse than it had led the public to believe (House of Lords Select Committee 2000; Millstone and van Zwanenberg 2000). Similarly, a national laboratory in the USA in the 1990s was found to have concealed, for more than a decade, the fact that tritium was leaking from a research reactor pool into the groundwater underneath Long Island in New York state. When the situation was discovered, laboratory officials continued to insist that there was no health or safety issue, despite great public unrest and protest (Lynch 2001). The continuing insistence by a number of countries that HIV infection is either absent or exists at low rates, while ignoring compelling evidence to the contrary, can be seen as one form of one-way communication that is intended to promote only the 'good news' and prevent an accurate analysis of a looming or existing public health crisis.

Many reporters and citizen watchdog groups are wary of the public information approach to PR practice. As there is little or no attempt to interact with interested parties outside the organisation except to provide information, these parties often are suspicious of the motives of the organisation – too often, rightly so. In many instances, media and public requests for information from the institution are required to go through the PIO, or only the PIO is allowed to talk to members of the public or press on behalf of the institution. For a science reporter bent on interviewing a scientist about a new scientific finding, this tactic seems an arbitrary barrier to reporting a solid story; for members of activist groups who may be critical of the organisation (such as a university engaged in animal research), the PIO is viewed as an obstacle designed to 'protect' the university and effectively hide 'what's really going on' from public scrutiny (Nelkin 1995).

It should be noted here that the explanatory (public information) approach to PR is still basically a one-way street, or at least highly asymmetrical. While the organisation and its PR practitioners may part with more information than those who practise simple press agentry, there still is little or no feedback from the public. Explanatory PR may employ focus groups, polls and surveys, and other means of finding out what the public knows or thinks in order to determine the right 'spin', but it does not engage in any two-way dialogue with its publics (Grunig and Grunig 1992).

Just as press agentry did, the practice of one-way, explanatory PR has failed science, as it has most other fields. Both have helped to foster a wary and mistrustful public, and this mistrust engenders self-fulfilling institutional paranoia on the part of the scientific enterprise, as noted by Bauer et al. (2007):

> Mistrust on the part of scientific actors is returned in kind by the public. Negative public attitudes, revealed in large-scale surveys, confirm the assumptions of scientists: a deficient public is not to be trusted.

Yet asymmetrical communication models are still the preferred mode of PR practice for many scientific organisations, with the public information model probably most widely practised in the scientific world today. This practice is best demonstrated by the increasing numbers of science writers being recruited by universities and research agencies, often to write lay-language brochures, news materials, website material and annual reports. While such practices have been a staple of US science PR activity for quite some time, public information work like this is relatively new in much of Europe and the developing countries (Moore 2000).

Nelkin (1995) does an excellent job of limning the history of PR in the USA as it applies to the scientific enterprise, tracing the lineage through the Smithsonian Institution's identification, as early as 1847, of the press as a means to inform the public about science, to the formation of the Science Service in 1930 by Edwin Scripps as a vehicle for promotion of favourable public images of science, to a host of public information activities undertaken by medical societies and other institutions to 'get the word out' about science. This cause took on greater urgency in the post-*Sputnik* era, as scientists increasingly bemoaned their perception of a growing deficit in scientific knowledge on the part of the public. More recently, Nelkin explains, PR at the corporate level has turned to 'damage control' as an essential function of preserving corporate credibility and enhancing company claims.

Much of the research in this realm, however, once again focuses on the direct interactions between scientist and reporter, sometimes including the role of the science information officer or equivalent, but only at the programme or function level (Weigold 2001).

Shifting to symmetrical communication?

While many scientific organisations may say they are doing it, or may even believe they are doing it, few actually encourage or engage in true dialogue with the public or publics, an approach that is termed symmetrical communication. Efforts at public engagement in science may be seen as examples of a commitment to two-way symmetrical communication. In a comprehensive survey of hundreds of organisations and their approaches to PR, Jim and Larissa Grunig argue that two-way symmetrical communication produces better long-term relationships with the public than do asymmetrical approaches to PR (Grunig and Grunig 2006).

The goal of two-way symmetrical communication is mutual satisfaction of both parties, the scientific organisation and its publics, with the relationships that exist between them. As the Public Relations Society of America affirms in a statement adopted by its governing body in 1988: 'Public relations helps an organization and its publics mutually adapt to each other.'

The mutual-satisfaction, or symmetrical, approach to PR emphasises true interaction between organisations and their publics. This approach requires a commitment to transparency on the part of the organisation; negotiation, compromise and mutual accommodation; and institutionalised mechanisms of hearing from and responding to the public. It places a premium on long-term relationship-building with all the organisation's strategic publics: taxpayers, media, shareholders, regulators, community leaders, donors and others (Grunig 1997).

Moreover, a variety of new technologies are available to make symmetrical communication possible – and amenable to research scrutiny. The World Wide Web provides a number of platforms, from online discussion groups to chat forums to blogs, that allow real-time, person-to-person communication with members of the public valuable to an institution. But the commitment to symmetrical communication falls short if the organisation hears, but does not respond to, the concerns or issues of its public. Mutual satisfaction mandates that organisations be open to reasonable changes requested of them – just as effective public engagement programmes in science should signal a willingness to incorporate public input in science policy or regulatory initiatives. Very little research attention has been paid to the direction or philosophy of communication flow from the managers of scientific institutions.

For some communication tasks, scientific organisations may be perfectly justified in using asymmetrical communication models. Filling an auditorium for an important lecture by a Nobel laureate is a publicity job, pure and simple. Preparing brochures and articles that clearly and simply articulate the research conducted or promoted by the organisation is explanatory PR at its best. It is, however, the rare scientific organisation that devotes significant resources to meaningful symmetrical communication (Nelkin 1995; Gregory and Miller 1998).

Managing the trust portfolio

One of the results of asymmetrical communication with the public is that public trust in the scientific enterprise is not robust, and may even have begun to drop in recent decades, although measures of this variable are not easy to come by historically (cf. Royal Society 1985; Haerlin and Parr 1999; House of Lords Select Committee 2000; AAAS 2004). At the Genetics and Public Policy Center in Johns Hopkins University (Washington, DC, USA), we have noted increasing concern about scientists 'playing God' and being unwilling or unable to resist the temptation to do things that are 'wrong' for money or prestige (Kalfoglou et al. 2004).

In this context, it might be useful for both practitioners and communications scholars to think of the PR function in a scientific organisation in terms of managing the institution's trust portfolio. By the trust portfolio, I mean the principal relationships that exist between the organisation and its many stakeholders. Science PR, done effectively and strategically, is the primary tool for managing this portfolio, and helps the other parts of the organisation to do their job more effectively by cultivating or maintaining trust in the ability of the organisation to do science, advocacy or science policy.

For example, public affairs officers (PAOs, yet another term for PR practitioners, most often used in government) at a government-funded research institution may have a number of stakeholders for whom science communication would be helpful in establishing and maintaining trust. First and foremost, the organisation is probably concerned about its funding stream, and the appropriate kinds of PR can help the agency or laboratory convince legislators or agency heads that money sent to this organisation is money well spent, and that its research is top quality and worth supporting. Second, the organisation probably has a need to make sure that other scientists and researchers elsewhere know about the range of research being conducted there, in order to facilitate collaboration and keep abreast of scientific research conducted by other organisations, and to position the organisation as a credible and reliable scientific collaborator (or future employer). Third, the organisation may need to have a good relationship with the people in the area surrounding the facility: government-run laboratories increasingly face the need to maintain the trust and support of their local communities in order to do research in community settings.

Of course, media are an important public, but in reality the organisation is seldom interested in the media for the media's sake. It is interested in media because they are able to reach other, primary stakeholders whose actions directly affect the ability of the organisation to stay open and conduct research. Media in this context are third-party validators; as Nelkin (1995) posits: 'Scientists ventriloquate through the media to those who control their funds'. Bad press certainly can affect the disposition of key stakeholders toward the organisation, and conversely, good press can validate the work and integrity of the organisation among groups that materially affect the organisation's ability to do its research. Good PR practitioners, however, never confuse the route they use to get to strategic publics with the publics themselves.

Similarly, good communications researchers should not focus unduly on the role of media in science communications. Media studies are low-hanging fruit from many perspectives – discrete, quantitative, archivable – yet they give only a very incomplete picture of trust relationships between an organisation and its true stakeholders. As Gregory and Miller (1998) observe:

> Key to the relationship between science and the public is trust. This trust is established through the negotiation of a mutual understanding, rather than through statements of authority or facts. Responsibility for the trust vested in the scientific community rests both with the institutions of science and with each individual member.

Components of the trust portfolio

Trust is not a single factor; it is actually an amalgam of several independent variables, all of which contribute to a robust and successful trust portfolio (Grunig and Hung 2002). Three important components of trust are:

- *Competence* – can the organisation do the work that is expected of it? Do the researchers who work there have the right credentials, and are they considered pre-eminent in their field? If the organisation is also a teaching organisation, like a university, can the researchers successfully integrate their research with teaching and turn out excellent young scientists? If it is a research unit in a hospital, can the researchers identify new therapies or innovative treatments that improve the quality of healthcare at the hospital?
- *Integrity* – do outside observers believe that researchers and management at the organisation know the difference between right and wrong in a science setting? Do they know the safety and ethical codes of conduct for medical research? Would the organisation know it should halt research that was found to have harmful environment effects? Do the researchers take short cuts to get their research published, or fabricate their data? Would they know better than to hide data that did not fit their hypothesis or match the goals of their funding agency?
- *Dependability* – do researchers who know the 'right thing to do' always do it? If faced with the chance to publish quickly by omitting a few steps in the research process, would they still follow the full protocol? If it is possible to make money or boost their recognition by failing to disclose negative data or an experiment gone awry, would they always still disclose it? Can political or managerial pressure make the best intentioned scientists opt to do the expedient thing, and not necessarily the right thing?

Most science organisations use PR to focus only on the first variable, competence. They seek media stories that illustrate successful research (in the case of research-performing institutions), or point out scientific issues consistent with the organisation's point of view (for health or science advocacy organisations, for example).

But competence is only one part of trust, and possibly the part least valued by external stakeholders. Nevertheless, competence and credibility are disproportionately – indeed, almost exclusively – reflected in scholarly studies of trust in science communication (Weigold 2001; Bauer et al. 2007).

Organisational goals of public relations

Most people – including communication scholars, if one were to judge by the literature – think of PR as something that one office in an organisation does. In reality, scientific institutions that are effective in their job of engendering trust and developing satisfying relationships with their publics must be effective at four different levels of organisational management. Certainly, communications success at the lower levels contributes to success at the higher levels. But an institution has little hope of effectively using PR until and unless it also is committed to, and attains, success at higher organisational levels.

The *programme level* is the individual component of an overall PR programme, such as media relations, publications, events planning and so on. The effectiveness of these individual programmes usually can be deduced by whether they meet specific objectives: do they change knowledge, attitudes or behaviours of the publics to which they are targeted? But success at the programme level of organisation does not guarantee, or necessarily even contribute to, success of the institution as a whole unless the publics reached are the ones that really matter to the health and survival of the institution, and such success as a programme contributes to cultivation of mutually satisfactory relationships with these strategic publics. A media relations programme that consistently produces reams of coverage in national newspapers about an institution's research is disconnected from the overall effectiveness of the institution if the strategic publics – policy-makers, say, or government funding agencies – do not read those newspapers, or if these publics use other criteria to judge an organisation's success.

When communication scholars study science PR, it is almost always this programme level – the tension between the science information officer and publics such as the media or policy-makers – that is the object of research or attention (Friedman et al. 1986; Weigold 2001).

The *functional level* is the overall communications or PR function of the institution, typically including all of the individual programme level units discussed above. While the PR function of an institution may appear to be effective – communications teams may regularly win awards for excellence in news writing or campaign planning – its effectiveness is only as good as its relationship to overall institutional management objectives. For example, if the PR function of a biomedical research institute does not understand that donors and foundations are also a critical element of the PR portfolio, the PR function cannot be considered successful.

At the *organisational level*, PR must contribute in some way to the organisation's bottom line – financially, in the case of a corporate entity; by attracting new students or helping retain world-class faculty at a university; or by attracting membership or donor support for non-profit advocacy organisations. Public relations is most effective at the organisational level when it helps the organisation to identify its strategic publics, how best to interact with them, and their expectations in return. This is a management function of PR, and requires that PR has a place 'at the table' among senior organisation executives to be truly successful.

Finally, at the *societal level*, PR professionals can help their organisations understand what it means to be socially responsible and help contribute to the ethical behaviour and social commitment of the organisation. At this level, management of the trust portfolio goes beyond the trust engendered between the organisation *per se* and its publics; it helps the organisation manage the trust portfolio for the entire scientific enterprise. Socially responsible scientific organisations help cultivate public trust in science and technology; PR – if empowered by management – can play a vital role in articulating social responsibility and finding ways for an organisation to allay public mistrust in, and wariness of, science and scientists.

Understanding the role of PR in science communications will require a much more robust research effort aimed at elucidating these organisational and societal level PR activities. Scholars need to examine, for example, the degree of disconnection between PR practices at the programme or function level (which, in my experience, are most likely to exhibit tendencies toward symmetrical communication) and the organisational level. Activities at higher management levels frequently undermine programme-level attempts at symmetry by putting in place rigid information control policies for corporate managers.

Evaluating public relations activities of research organisations

It has often been observed that, for people whose lives revolve around data, scientists are all too willing to let their PR activities be guided by chance comments from a programme funding officer or the cocktail chatter of an elected official. In fact, PR success can be measured, qualitatively and quantitatively, and should be built on solid formative and evaluative research (see Chapter 17 in this volume). Even when otherwise excellent PR departments are asked to provide metrics about their overall contributions to the success of their organisations, they too often respond – and are not challenged to do better – by providing evidence of success only at the programme level, such as number of hits on a university website, or number of press releases sent out that made their way into public news coverage.

Public relations should be informed by the same rigour and hypothesis-driven research as the best scientific experiment. Formative research is used to identify strategic publics and the ways those publics acquire new information or new ideas. This could include surveys, focus groups, one-on-one interviews with key members of a strategic public, or even content analysis of web postings in chat rooms about an organisation or its issues. Evaluative research should test the effectiveness of PR messages, approaches or techniques – ideally before and after specific communication initiatives – to see how well they have helped to align an organisation better with its strategic publics (Grunig and Grunig 2001). In principle, all PR activities should be grounded in the relevant communications theories and informed by intimate familiarity with the communications literature. Unfortunately, few PR practitioners in science have received training in communication theory, and fewer still actively keep up with the literature.

Quantitative research probably is the type of research understood best, both by communications scholars and by managers in a scientific organisation, and there are times when a well designed, scientifically meaningful survey of key stakeholders provides the best data from which to plan or evaluate a PR programme. But good qualitative research – perhaps including focus groups with informed and forthright stakeholders, or community-leader interviews – can yield excellent analysis, especially early in a communications campaign.

Good PR practitioners also distinguish between measures of process and measures of outcomes. Too often, PR departments offer process measures as testimony to their effectiveness: the number of news releases sent out, media calls per month, the number of new employees who sit through a PR unit during orientation. These process measures, however, are virtually meaningless unless they are tied to outcomes: short- and long-term impacts on cognitions, attitudes and behaviours of strategic publics (Grunig and Grunig 2001). Process measures also offer limited insight into corporate trust issues, unless tied to the same outcomes.

In science, too, trust starts with the CEO

Public relations is a strategic function of a successful organisation, as well as a tactical one. Too many scientific organisations see only the tactical value of PR, and are content to manage it through the human resources department, through laboratory administration, or through some other programme not at all connected functionally to PR.

This has important ramifications for the qualifications of PR managers in scientific organisations. They really must understand the scientific issues that come before the senior management, and speak with authority to the scientists and researchers at the lab bench. They must be seen as independent experts in their own right – there should be no temptation on the part of the CEO or other senior staff to make PR decisions without them. And they must have the full and absolute backing of senior management in implementing practices of engagement with stakeholder publics, but also be seen as credible by those stakeholders.

These internal relationships have been very poorly studied in science organisations, although they undoubtedly have great impact on science communications as an enterprise.

Because PR is a function of entire organisations, not just science communicators or science officers, unless science communications researchers turn their attention to this critical component of PR practice, and to its impact on the trust portfolio at the societal level, we will have great difficulty in comprehending the contributions of PR to the public understanding of science and technology.

Suggested further reading

Grunig, J. E. and Grunig, L. A. (2001) *Guidelines for Formative and Evaluative Research in Public Affairs: A Report for the Department of Energy Office of Science*, College Park, MD: Department of Communication, University of Maryland; www.instituteforpr.org/files/uploads/2001_PA_Research.pdf

Nelkin, D. (1995) *Selling Science: How the Press Covers Science and Technology* (revised edn), New York: W.H. Freeman & Co.
Weigold, M. F. (2001) 'Communicating science: a review of the literature', *Science Communication*, 23: 164–93.

Other references

AAAS (2004) *AAAS Survey Report*, Washington, DC: American Association for the Advancement of Science. www.aaas.org/news/releases/2004/aaas_survey_report.pdf
Bauer, M. W., Allen N. and Miller S. (2007) 'What can we learn from 25 years of PUS survey research? Liberating and expanding the agenda', *Public Understanding of Science*, 16: 79–95.
Bernard, K. and Duffy, S. (1982) 'Persuasion and uplift in American theatrical advertising during the Depression', *Journal of American Culture*, 5: 66–71.
Friedman, S., Dunwoody, S. and Rogers, C. L. (eds) (1986) *Scientists and Journalists: Reporting Science as News*, New York: Free Press.
Gregory, J. and Miller, S. (1998) *Science in Public: Communication, Culture, and Credibility*, New York: Plenum Press.
Grunig, J. E. and Grunig, L. A. (2001) *Guidelines for Formative and Evaluative Research in Public Affairs: A Report for the Department of Energy Office of Science*, College Park, MD: Department of Communication, University of Maryland; www.instituteforpr.org/files/uploads/2001_PA_Research.pdf
—— (2006) 'Characteristics of excellent communication,' in Gillis, T. L. (ed.) *The IABC Handbook of Organizational Communication: A Guide to Internal Communication, Public Relations, Marketing, and Leadership*, San Francisco, CA: Jossey-Bass, 3–18.
Grunig, J. E. and Hung, C. F. (2002) 'The effect of relationships on reputation and reputation on relationships: a cognitive, behavioral study', paper presented at the *PRSA Educator's Academy 5th Annual International, Interdisciplinary Public Relations Research Conference*, Miami, Florida, 8–10 March 2002.
Grunig, J. E. and Hunt, T. (1984) *Managing Public Relations*, San Diego, CA: Holt, Rhinehart & Winston.
Grunig, L. (1997) 'Excellence in public relations', in Caywood, C. L. (ed.) *The Handbook of Strategic Public Relations and Integrated Communications*, New York: McGraw-Hill, 286–300.
Grunig, L., Grunig, J. E. and Dozier, D. M. (2002) *Excellent Public Relations and Effective Organizations: A Study of Communication Management in Three Countries*, Mahwah, NJ: Lawrence Erlbaum Associates.
Haerlin, B. and Parr, D. (1999) 'How to restore public trust in science', *Nature* 400: 499.
House of Lords Select Committee on Science and Technology (2000) *Science and Society, 3rd Report*, London: HMSO.
Kalfoglou, A., Scott, J. and Hudson, K. (2004) *Reproductive Genetic Testing: What America Thinks*, Washington, DC: Genetics and Public Policy Center.
Lynch, M. (2001) 'Managing the trust portfolio', in *Proceedings of the PCST2001 Conference*. Public Communication of Science and Technology network. http://visits.web.cern.ch/visits/pcst2001/proceedings_list.html
Marchand, R. (1998) *Creating the Corporate Soul: The Rise of Public Relations and Corporate Imagery in American Public Relations*, Berkeley, CA: University of California Press.
Millstone, E. and van Zwanenberg, P. (2000) 'A crisis of trust: for science, scientists or for institutions?', *Nature Medicine*, 6: 1307–8.
Moore, A. (2000) 'Would you buy a tomato from this man? How to overcome public mistrust in scientific advances', *EMBO Reports*, 1: 210–12.
Royal Society (1985) *The Public Understanding of Science*, London: Royal Society.

11

Environmental action groups and other NGOs as communicators of science

Steven Yearley

Introduction: the elective affinity of environmental action groups and NGOs with science communication

In conjunction with its report on the human and environmental impacts of man-made chemicals (Greenpeace 2003), in 2003 British Greenpeace ran a campaign advertisement in national newspapers showing a figure closely resembling Michelangelo's *David*, complete with its small genital endowment. The accompanying text suggested that people should begin to worry about threats to men's reproductive capacity owing to the environmental release of hormone-mimicking substances. Chemicals used in plasticisers and other applications could be 'feminising' the environment and leading to declining male fertility, in humans and in wild animals too. This advertisement was indicative of Greenpeace's strategy. It expressed in an arresting way the supposed facts of the case – here was a new form of harm arising from a novel and unanticipated form of environmental pollution – but it was also significantly misleading, as the chemicals were unlikely, on anyone's view, to lead to a threat to the size of male members. This advertisement encapsulated a key challenge in the public communication strategy of environmental non-governmental organisations (NGOs): the need to balance powerful, evocative images with the perceived demands of accuracy.

This chapter examines the increasing importance of environmental NGOs as mediators of scientific information in policy and other public arenas, and the challenges they face in positioning themselves in relation to shifting scientific orthodoxies. Environmental campaign organisations have been important in supplying arguments about, and publicising problems in relation to, a very large number of environmental issues. Rather than trying to conduct a review of these issues, this chapter focuses first on one leading example, climate change; derives some points of principle from it, then assesses their generalisability by applying them to a contrasting case – genetically modified organisms.

Opponents of Greenpeace and other environmental organisations have frequently criticised them for favouring the slick image over the accurate message, but this criticism – although interesting and important – implicitly acknowledges something even more interesting. The key point is that environmental pressure groups can be called to account on this issue precisely because the persuasive power of their message depends on the notion that their claims have a basis in factual accuracy, that they are not matters of opinion. Environmental organisations, more than any other type of campaigning group, need to persuade the public that things are in fact the way they say things are, even when some of the claims they are making seem – at first glance at least – to be counter-intuitive or implausible: that plastics can make you infertile, or that burning fossil fuels can unsettle the entire global climate.

Thus, in what is clearly today's pre-eminent environmental debate, environmentalists are keen to assert that global warming is in fact taking place, and that humanly caused changes in the make-up of the atmosphere are responsible. Central to the campaign is the claim that, in fact, humans are causing global warming and that, in fact, warming will have specific adverse implications.

Indirect testimony to this factual and scientific orientation is given in the following observation about Friends of the Earth (FoE).[1] To celebrate its 21st birthday, the environmental organisation published a celebratory booklet (FoE 1992). With a large supporter base, active local groups, regular coverage in influential media, strong campaign teams and widespread name recognition, the organisation had a lot to boast about. 'Yet the item chosen to begin this celebratory publication, immediately after the contents page, was a quote from a leading environmental journalist praising the group as a "reliable and indispensable source of information"; this was followed by a comment from the head of Her Majesty's Inspectorate of Pollution [the forerunner to Britain's Environment Agency] lauding the quality of its "technical dialogue"' (Yearley 2005a: 113). Given all the things that could have been chosen to feature at the start of FoE's anniversary document, this selection was surprising and telling.

In short, I suggest that there is an elective affinity between environmental campaign organisations and scientific claims that is, to a large degree, distinctive among pressure groups. This gives environmental campaign groups an urgent interest in science communication issues and makes them significant science communication actors.

Climate change as an emblematic science communication challenge for NGOs

Scientists have been aware for over a century that the climate undergoes significant variation, and there has long been a concern that human society could not for ever count on a stable climate. As climate research was refined, largely thanks to the growth in computer power in the 1970s and 1980s, the majority opinion endorsed the earlier suggestion that enhanced warming, driven by the build-up of atmospheric carbon dioxide, was the likeliest problem to face humankind in the short to medium term. Environmental groups are reported to have been initially wary of campaigning around this issue (Pearce 1991: 284), as it seemed such a long shot, and with such high stakes. With acid rain on the agenda, and many governments active in denying scientific claims about even this comparatively straightforward effect, the stakes seemed too high in the 1980s to declare publicly that emissions might be sending the whole climate out of control. Worse still from a campaigner's point of view, at a time when environmentalists were looking for concrete successes, the issue seemed almost designed to provoke and sustain controversy. The records of past temperatures, and particularly of past atmospheric compositions, were often not good, and there was the danger that rising trends in urban air-temperature measurements were simply an artefact: cities had simply become warmer as they grew in size. The heat radiating from the Sun is known to fluctuate, so there was no guarantee that any warming was a terrestrial phenomenon due to 'pollution' or other human activities. Others doubted that additional carbon dioxide releases would lead to a build-up of the gas in the atmosphere, as the great majority of carbon is in soils, trees and the oceans, so sea creatures and plants might simply sequester more carbon. And even if the scientific community was correct about the build-up of carbon dioxide in the atmosphere, it was fiendishly difficult to work out what the implications of this would be in order to build campaigns with local resonance.

Hart and Victor (1993) tracked the interaction between climate science and US climate policy from the 1950s to the mid-1970s, by which time greenhouse emissions had begun to be 'positioned as an issue of pollution' (ibid. 668); the climate, 'scientific leaders discovered, could be portrayed as a natural resource that needed to be defended from the onslaught of industrialism' (ibid. 667). Subsequently, according to Bodansky (1994: 48), the topic's rise to policy prominence was assisted by other considerations. There was, for example, the announcement of the discovery of the 'ozone hole' in 1987; this lent credibility to the idea that the atmosphere was vulnerable to environmental degradation and that humans could unwittingly cause harm at a global level. Also important was the coincidence in 1988 between Senate hearings into the issue and a very hot, dry summer in the USA. Nonetheless, most politicians responded to the warnings in the 1980s with a call for more research.

One significant outcome of this support for research was the setting up in 1988 of a new form of scientific organisation, the Intergovernmental Panel on Climate Change (IPCC) under the aegis of the World Meteorological Organization and the United Nations Environment Programme. The aim of the IPCC was to bring together the leading figures in all aspects of climate change, with a view to establishing in an authoritative way the nature and scale of the problem and identifying candidate policy responses. This initiative was accorded significant political authority and was novel in significant ways. Among its innovations were the explicit inclusion of social and economic analyses alongside the atmospheric science, and the involvement of governmental representatives in the agreeing and authoring of report summaries. 'While by no means the first to involve scientists in an advisory role at the international level, the IPCC process has been the most extensive and influential effort so far' (Boehmer-Christiansen 1994: 195).

As is widely known, the IPCC has met with determined criticism. At one end, there have been scholars and moderate critics who have concerns about the danger that the IPCC procedure tends to marginalise dissenting voices, and that particular policy proposals (such as the IPCC-supported Kyoto Protocol) are maybe not as wise or as cost-effective as proponents suggest (e.g. Boehmer-Christiansen and Kellow 2002; Boehmer-Christiansen 2003). There are also very many consultants backed by the fossil-fuel industry who are employed to throw doubt on claims about climate change. Freudenburg (2000) offers a discussion of the social construction of 'non-problems'. These claims-makers have entered into alliance with right-leaning politicians and commentators to combat particular regulatory moves, as detailed by McCright and Dunlap (2000, 2003). Informal networks, often web-based, have been set up to allow 'climate-change sceptics' to exchange information, and they have welcomed all manner of contributors, whether direct enemies of the Kyoto Protocol or more distant allies, such as opponents of wind farms or conspiracy theorists who see climate-change warnings as the machinations of the nuclear industry.

Gifted cultural players, including Rush Limbaugh and Michael Crichton, have waded into this controversy; Crichton's (2004) novel *State of Fear* includes a technical appendix and author's message on the errors in climate science. In his book, Crichton even goes as far as to offer his own estimate of the rate of global warming – 0.812436 degrees for the warming over the next century (Crichton 2004: 677). Crichton and others have concentrated not only on the scientific conclusions (and their disagreements with them), but have looked at putative explanations for the persistence of error in 'establishment' science and much of the media, to which I shall return shortly. At the same time, mainstream environmental NGOs have tended to argue simply that one should take the scientists' word for the reality of climate change, a strategy about which they have clearly been less enthusiastic in other cases (Yearley 1993: 68–9).

The rhetorical difficulties of doing this were already foreshadowed in the strategy of FoE in London over 15 years ago: in 1990, campaign staff working on climate change issues were disturbed by a programme aired on the UK's Channel 4 television, in the *Equinox* series, that sought to question the scientific evidence for greenhouse warming. The programme even implied that scientists might be attracted to make extreme and sensational claims about the urgency of the problem in order to maximise their chances of receiving research funding. The programme was criticised in the 'Campaign News' section of the FoE magazine *Earth Matters*. An unfavourable comparison was drawn between the sceptical views expressed in the programme and the conclusions of the IPCC, with whose scientific analysis FoE was generally in agreement. The FoE article invoked the weight of 'over 300 scientists [who] prepared the IPCC's Science Report compared to about a dozen who were interviewed for Equinox'.[2] When scientists with apparently good credentials are seen to disagree, it is very difficult for environmentalists to take the line that they are simply in the right. It seems a reasonable alternative to invoke the power of the majority. But this remedy cannot always be adopted – in many areas where environmentalists believe themselves to be factually correct, they have been in the scientific minority, at least initially. In March 2007, Channel 4 repeated this attention-seeking strategy, broadcasting a programme unambiguously entitled *The Great Global Warming Swindle*. The response of NGOs and 'green' commentators was essentially the same: we should trust the advice of the great majority of well-qualified scientists who accept the evidence of climate change. Environmental groups looked to invoke the possible vested interests of the critics in order to make sense of the programme-makers' and contributors' continued scepticism.

In the relationship between the IPCC – indeed, the whole climate change-regulation community – and its critics, not only the science, but the various ways in which the science is legitimated, have come under attack (Lahsen 2005). Critics have been quick to point to the supposed vested interests of this community. Its access to money depends on the severity of the potential harms that it warns about; hence – or so it is argued – it inevitably has a structural temptation to exaggerate those harms. As it was working in such a multi-disciplinary area, and with high stakes attached to its policy proposals, the IPCC attempted to extend its network widely enough so as to include all the relevant scientific authorities; it was important that the IPCC should not be dominated by meteorologists or atmospheric chemists. But this meant that the IPCC ran into problems with peer reviewing and perceived impartiality; there were virtually no 'peers' who were not already within the IPCC (for an analysis of the accusations that could be levelled on this basis, see Edwards and Schneider 2001). Conventional peer reviewing relies on there being few authors and many (more-or-less disinterested) peers; the IPCC effectively reversed this situation. When just one chapter in the 2001 Third Assessment Report has 10 lead authors and over 140 contributing authors,[3] then it is clear that this departs from the standard notions of scientific quality control.

If challenged, the IPCC tended to fall back in line with the classic script of 'science for policy' (Yearley 2005b: 160–2); the IPCC legitimated itself in terms of the scientific objectivity and impartiality of its members. But critics were able to point out that the scientific careers of the whole climate-change 'orthodoxy' depend on the correctness of the underlying assumptions. Worse, the IPCC itself selects who is in the club of qualified experts, and thus threatens to be a self-perpetuating elite community (this line of attack is described by Boehmer-Christiansen 1994: 198). This was exactly the point that Crichton picked up. Many of his speeches and articles are available on his website, alongside a very specific demand that the work not be reproduced (www.michaelcrichton.com/books-stateoffear.html). Without therefore quoting him – which he forbids – his principal argument is that the key requirements are a form of independent verification for claims about climate change, and the guarantee of access to unbiased information. However well meant, this is clearly an unrealistic demand, as there is no-one with scientific skills in this area who could plausibly claim to be entirely disinterested. There is no Archimedean point to which to retreat, and environmental groups will correctly claim that such demands for a review are primarily ways to put off taking action. Crichton further muddies the water by proposing to offer his own estimate of future climate change to six decimal places; although the ridiculous precision clearly signals some jocular intent, the idea that even he (a medical doctor turned author) can offer a temperature-change forecast implies that there are lots of people able to make independent judgements. In fact, there are relatively few, and a central science communication challenge for environmental groups is to distinguish between those who can credibly comment and those who cannot.

Although they have found it hard to participate in the central scientific debate, and have been obliged to take up the (for them) unusual position of defending the correctness of mainstream science, environmental action groups have found other activities that they have been able to pursue. For example, in the USA they have been active in trying to identify novel ways to press the government to change its position on climate change aside from simply bolstering the persuasiveness of climate science and trying to rebut the claims of critics such as Crichton. Thus in 2006 the

Center for Biological Diversity (CBD), the Natural Resources Defense Council and Greenpeace learned that their inventive use of the Endangered Species Act to sue the US government for protection of polar bears and their habitat in Alaska had won concessions from the government. In its campaigning, the CBD had argued that oil exploration in the far north would harm polar bears and their hunting grounds, but they also suggested that ice melting caused by global warming was responsible for additional habitat loss and harm to bears that need large expanses of solid ice in spring for successful hunting.[4] Potentially, the endangered species legislation could force the government to examine the impact on polar bears of all actions in the USA (such as energy policy), not just activities local to polar bear habitat.

In this instance, environmental NGOs have been stuck in a dilemma. What they see as the world's leading environmental problem is fully endorsed by the mainstream scientific community. In January 2004, the UK government's chief scientific adviser, Sir David King, gave his judgement that climate change posed a greater threat than terrorism.[5] The NGOs' principal efforts have accordingly been directed at restating and emphasising official findings, finding novel ways to publicise the message, and countering the claims of greenhouse-sceptics. The difficult part of the dilemma is that such statements in favour of the objectivity of the scientific establishment's views mean that it is harder to distance themselves from scientists' conclusions on other occasions without appearing arbitrary or tendentious.

Environmental organisations and GMOs: communicating safety and risks

The case of genetically modified (or genetically engineered) organisms was just the opposite of climate change in the sense that environmental groups were, initially at least, out of line with the views of the scientific establishment; the science communication issues were accordingly very different. In this case, the principal issues addressed safety and safety-testing. Here was a new product, whether GM crop, animal or bacterium, that needed to be assessed for its implications for consumers and the natural environment. Of course, all major industrialised countries had some sort of procedures for testing new foodstuffs. But the leading question was how novel were GM products taken to be, and thus what sorts of tests they should be exposed to. For some, the potential for the GM entity to reproduce itself or to cross with living relatives in unpredictable ways suggested that this was an unprecedented form of innovation that needed unparalleled forms of caution and regulatory care. On the other hand, industry representatives and many scientists and commentators claimed that it was far from unprecedented. People had been introducing agricultural innovations for millennia by crossing animals, allowing animal-based 'sports' to flourish, and so on. Modern (although conventional) plant breeding already used extraordinary chemical and physical procedures to stimulate mutations that might turn out to be beneficial. On this view, regulatory agencies were well prepared for handling innovations in living, reproductive entities (Jasanoff 2005).

In this case, the strategy of environmental action groups was much more typical of their approach in the years since the late 1960s. They argued that the regulatory system was insufficiently demanding, and that the consequences of new technologies were not being examined closely enough. They suggested that governments, keen to promote economic success and to support agribusiness and the farm sector, were not taking enough care for consumers and the environment. Indeed, the protest over GM chiefly differed from preceding environmental controversies in just two ways. First, the GM debate closely combined worries over environmental impacts and over the health consequences of the new technology; second, the GM controversy in the late 1990s came after a period in which there had been growing cooperation between environmental groups and official bodies, with them often agreeing to work together on projects aimed at so-called sustainable development.

Despite the intensity of recent public controversies over genetic foodstuffs, work on genetic engineering has a three decade-long history. Environmental action groups were preparing their arguments before the main range of products came to market in the 1990s. In the USA, these products passed tests set by the government agencies (Food and Drug Administration, US Department of Agriculture, Environmental Protection Agency) relatively quickly. The question for environmental action groups was how to express doubts about the advisability of this new technology. Opponents were worried about specific impacts – the possibility of adverse environmental impacts and conceivable food safety issues – but they also had serious concerns over the potential direct intervention in nature that the technology (at least in principle) offered.

As the debate unfolded, a number of specific issues came to be the focus of campaigns. These included, for example, the impact of GM crops on beneficial insects, the likely difficulty of organic growers in keeping their crops free of GM contamination, the possible effects of GM foods on people with allergies (as allergy-promoting aspects of crops might accidentally be crossed into formerly innocuous foodstuffs), and contentious evidence that GM crops might be less nutritious than existing crops in unexpected ways. At the same time, campaigners were aware that there was a danger in offering very specific objections to the new technology since, if these objections were successfully countered, opposition might begin to crumble. Campaigners were literally intransigent as they feared that any accommodation to the new techniques would open the world to GM. Moreover, even if GM agriculture might arguably be less bad for the environment than present-day intensive farming, there was still a worry that the GM route was a one-way journey to a new style of relationship with nature, as one could not readily imagine how GM 'contamination' could be undone (see the views captured in the study reported by Stirling and Mayer 1999).

Given the campaigning need to maintain the line against GM in Europe, environmental action groups fused their scientific communication with other strategies in largely opportunistic ways (Priest 2001). Thus a broad anti-GM coalition emerged, with groups engaged in anything from direct destruction of trial GM crops to detailed research work (examining, for example, just how far pollen from GM crops could travel) and everything in between. Those with distinctively environmental concerns were joined by anti-globalisation protesters and – particularly in France – by groups devoted to protecting the livelihoods and way of life of smallholders against large seed and agrichemical companies. In many of these activities, the groups were happy to let environmental and health anxieties overlap and intensify each other. Thus, in a well-known incident in England where a Greenpeace 'decontamination' raid on a farm growing GM crops went wrong and protestors were arrested, the activists were all dressed in protective suits as though simple exposure to GM vegetables might be harmful (Yearley 2005a: 172–4).

At the same time, the more professionalised campaigning organisations began to concentrate on publicising and analysing environmental harms. The scientific evidence for adverse health effects was contentious and not easy to campaign around, but there were more readily agreed mechanisms by which GM planting might be causing environmental harms. In Britain, this emphasis was further promoted as an unintended consequence of government policy. Wishing neither to accede to activists' demands to ban GM crops nor to try to impose GM agriculture against popular sentiment, the UK government opted to put off the decision about GM cultivation by organising a series of field-scale trials over several years, aimed at investigating what the results of GM agricultural practice would be on wild plants, bird species, insects and so on. This strategy neatly provided a new target for sabotage attempts, but also focused the debate around impacts on the rural environment and away from impacts on consumers. Official countryside-protection agencies and more establishment conservation groups were also keen to see such tests done. The key irony here is that, given that today's GM food crops work either by killing off pests or through allowing weeds to be controlled more easily, there was a good chance that – even if GM crops behaved exactly as predicted – they would have a negative impact on wildlife. As there would be fewer weeds, there would be fewer seeds and insects, and thus less to sustain wild birds. Naturally, it was hard to believe that the decline in field weeds was the prime concern of individual consumers who worried about whether to buy GM foods, but it provided a reasonably objective basis for claiming that GM agriculture would have a negative impact on the British countryside and thus a sound legitimation for an anti-GM stance.

As I have pointed out, in practice the debate over GMOs in Europe was distributed over a range of issues. Despite this, the fundamental issue in EU legal assessments of GM crops and foodstuffs was the question of risk: were these new crops more risky than existing ones?

This framing of the issue was the predominant one in North America too. Disturbed by protestors' success in raising public disquiet about GM products, US companies and allied politicians felt that European resistance to GM imports should be combated by appeals to the World Trade Organization (WTO). A formal complaint was lodged in 2003: the USA hoped to use the WTO to force open European markets to US farm imports and seed companies. The USA argued there was no scientific evidence of harm arising from GM food and crops, as these products had all passed proper regulatory hurdles both in the US system and within the EU. Activists' concern over this legal move was twofold. First, they anticipated (correctly as it turned out) that the WTO would largely rule in favour of the USA; even if this has little effect in Europe, owing to developed consumer resistance, it will discourage people in other parts of the world from trying to regulate against GM agriculture. Second, there was a concern over the basis for the ruling (Winickoff et al. 2005). The WTO does not itself engage in additional scientific research on the issue, but makes a judgement about what the relevant law is and whether that legal interpretation has been followed. The WTO decision-makers took a narrow view of the basis for the decision: it should be about risk assessment. Environmental action groups and academic authors had little success in opening up a debate around this issue and in convincing the WTO body of the shortcomings of a US-style risk assessment approach (Busch et al. 2004).

A final distinctive opportunity for science communication around GM arose in relation to the consultative exercises that were run in several EU states and elsewhere (such as New Zealand) as a way of trying to win legitimacy for, or even decide on, public policy in relation to this contested issue (Hansen, 2005). The British exercise 'GM Nation?' (Horlick-Jones et al. 2007; Chapters 14 and 17 in this volume) was another option adopted by the UK government around the same time as the field-scale trials, partly as a way of being seen to do something without actually yet deciding for or against GM agriculture. In the UK case, at least, environmental action groups put a lot of effort into encouraging people to participate in the public debating exercise, even though it was clear that many participants would be frustrated as the debate would inform – but not determine – national policy.

Participation, consultation and public engagement as fora for NGOs' science communication

As the 'GM Nation?' and related initiatives indicate, the scope for public participation in environmental policy has grown greatly in recent years. The exact rationale for such participation has differed from one context to another, ranging from a wish to give citizens a say in democratic decision-making, to the suggestion that citizens may have insights into their local environment that are unavailable to the customary scientific experts (Yearley 1999; Kasemir et al. 2003; Chapter 12 in this volume). For the purposes of this chapter, the key issue is the response of environmental action groups to such initiatives.

On the face of it, one would expect environmental groups to have an affinity with such moves. Social movements tend to thrive in democratic societies and to espouse democratic principles. And on many occasions they call for government to respond to the supposed 'will' of the people, for example over GM food in Europe. But, at the same time, these organisations are aware that not all environmental objectives are popular; nor are popular policies necessarily environmentally benign. In Britain, the Blair government's decision to allow official online petitions to be created on the Downing Street website led to an enormous response when a record number, well over a million people, expressed their opposition to road pricing in early 2007. Equally, one of the largest acts of popular opposition to Blair's government had earlier been the protest against automatic increases in car fuel duty. In both cases, popular action appeared to favour personal consumption against environmental objectives.

Environmental groups are thus reluctant to relinquish control over the policy agenda to these public consultation initiatives, in part because they fear that such exercises might be manipulated by government (or business), but also because they are concerned that people may not favour the best environmental option. In many respects, environmental groups are as loath to hand environmental policy over to the public as are governments, even though environmental groups are happy to laud the wisdom of the public when the public happens to favour the same objectives as them. Citizens are thus deemed wise about GM foods, but less so about wind turbines, and least of all in their devotion to car ownership and use.

Questions of public consultation and participation come to the fore chiefly in open, pluralistic societies. A contrasting situation confronts environmental action groups in more restrictive cultures, notably China, a country of enormous importance for global environmental politics. It is believed that China overtook the USA to become the largest emitter of carbon dioxide in 2006;[6] China also has a large commitment to GM agriculture, and there are currently plans to introduce GM rice, which would move the Chinese diet from a low-GM profile to one of the world's most intensive. Although the Chinese state has a large and reasonably well resourced environmental protection agency (SEPA), environmental campaigning groups are generally not welcomed. Just a few international NGOs operate in China, and there is the curious phenomenon of the government-organised non-governmental organisations (GONGOs, Yang 2005: 50). Although some education- and membership-based environmental organisations operate, many environmental activists are attracted to alternative methods for communicating their message.

Several recent analyses in the Chinese and other East Asian and South-East Asian context have focused on the role of online communication (Yang 2003). As Yang points out:

> For web-based ENGOs, the internet makes up for their lack of resources and helps to overcome some political constraints. While the restrictive regulations create barriers to registering an NGO, web-based groups can stake out an existence on the internet.
>
> (Yang 2005: 59)

Mol (2006) also draws attention to the role of the internet in China and Vietnam in circulating information and allowing people access to information that they would otherwise have had difficulty accessing.

The potential of the internet has also been explored in the industrialised North, and not only as a forum for debate as discussed above in relation to climate change. The internet has been used for the provision of technical information: for example, FoE in Britain beat the official Environment Agency to provide online map-based information about local chemical pollution. Members of the public could search for possible sources of hazard by entering their postcode. Having shamed the Environment Agency into improving its public information, FoE has now withdrawn its site. In the USA, a similar job is handled by the well-known Scorecard site (www.scorecard.org: Chapter 13 in this volume).

Concluding remarks

The central claim of this chapter has been that environmental action groups, more than most other political and reform movements, have been obliged to act as communicators of science and technology because empirical claims about the state of the natural environment are core to their message. Often they have had to carry out this communication in circumstances where they disagree with the orientation of large parts of the scientific and technological 'establishment', and they have developed tools of argumentation for tackling this job. More recently – particularly over climate change – they have had to devise a new strategy for bolstering the IPCC and other mainstream science. The internet has proved to be a rich resource for such communication, both because it can handle detailed information and because the user can 'personalise' it by entering geographical data (zip codes or postcodes). In China, and other countries where environmental action groups face limitations on their activities, the internet has become a particularly indispensable means of environmental communication.

Notes

1 More precisely, it is FoE (England, Wales and Northern Ireland) since FoE Scotland is a separate organisation.
2 There was no author given for this report in *Earth Matters*, Autumn/Winter 1990, 4.
3 My example is Chapter 2, 'Observed climate variability and change'.
4 According to the CBD website: '"Short of sending Dick Cheney to Alaska to personally club polar bear cubs to death, the administration could not have come up with a more environmentally destructive plan for endangered marine mammals," said Brendan Cummings, ocean program director of the Center. "Yet the administration did not even analyze, much less attempt to avoid, the impacts of oil development on endangered wildlife".' See www.biologicaldiversity.org/swcbd/press/off-shore-oil-07-02-2007.html
5 'US Climate Policy Bigger Threat to World than Terrorism' was the headline in the UK newspaper *The Independent* (9 January, 2004)
6 See the report of the Dutch MNP (Milieu- en Natuurplanbureau) (last consulted on 22 June, 2007) at: www.mnp.nl/en/dossiers/Climatechange/moreinfo/Chinanowno1inCO2emissionsUSAinsecondposition.html

Suggested further reading

Horlick-Jones, T., Walls, J., Rowe, G., Pidgeon, N. F., Poortinga, W., Murdock, G. and O'Riordan, T. (2007) *The GM Debate: Risk, Politics and Public Engagement*, London: Routledge.
McCright, A. M. and Dunlap, R. E. (2000) 'Challenging global warming as a social problem: an analysis of the conservative movement's counter-claims', *Social Problems*, 47: 499–522.
Mol, A. P. J. (2006) 'Environmental governance in the Information Age: the emergence of informational governance', *Environment and Planning*, C24: 497–514.
Winickoff, D., Jasanoff, S., Busch, L., Grove-White, R. and Wynne, B. (2005) 'Adjudicating the GM food wars: science, risk and democracy in world trade law', *Yale Journal of International Law*, 30: 81–123.
Yang, G. (2005) 'Environmental NGOs and institutional dynamics in China', *China Quarterly*, 181: 46–66.

Other references

Bodansky, D. (1994) 'Prologue to the Climate Change Convention,' in Mintzer, I. M. and Leonard, J. A. (eds), *Negotiating Climate Change: The Inside Story of the Rio Convention*, Cambridge: Cambridge University Press. 45–74.

Boehmer-Christiansen, S. (1994) 'Global climate protection policy: the limits of scientific advice, Part 2', *Global Environmental Change*, 4: 185–200.

—— (2003) 'Science, equity, and the war against carbon', *Science, Technology and Human Values*, 28: 69–92.

Boehmer-Christiansen, S. and Kellow, A. J. (2002) *International Environmental Policy: Interests and the Failure of the Kyoto Process*, Cheltenham: Edward Elgar.

Busch, L., Grove-White, R., Jasanoff, S., Winickoff, D. and Wynne, B. (2004) *Measures Affecting the Approval and Marketing of Biotech Products, Amicus Curiae* brief submitted to the Dispute Settlement Panel of the World Trade Organization.

Crichton, M. (2004) *State of Fear*, London: HarperCollins.

Edwards, P. N. and Schneider, S. H. (2001) 'Self-governance and peer review in science-for-policy: the case of the IPCC Second Assessment Report', in Miller, C. A. and Edwards, P. N. (eds), *Changing the Atmosphere: Expert Knowledge and Environmental Governance*, Cambridge, MA: MIT Press, 219–46.

FoE (1992) *Twenty-One Years of Friends of the Earth*, London: Friends of the Earth.

Freudenburg, W. R. (2000) 'Social constructions and social constrictions: toward analyzing the social construction of "the naturalized" as well as "the natural",' in Spaargaren, G., Mol, A. P. J. and Buttel, F. H. (eds), *Environment and Global Modernity*, London: Sage, 103–119.

Greenpeace UK (2003) *Human Impacts of Man-made Chemicals*, London: Greenpeace.

Hansen, J. (2005) 'Framing the public: three case studies in public participation in the governance of agricultural biotechnology', PhD thesis, Florence: European University Institute.

Hart, D. M. and Victor, D. G. (1993) 'Scientific elites and the making of US policy for climate change research', *Social Studies of Science*, 23: 643–80.

Horlick-Jones, T., Walls, J., Rowe, G., Pidgeon, N. F., Poortinga, W., Murdock, G. and O'Riordan, T. (2007) *The GM Debate: Risk, Politics and Public Engagement*, London: Routledge.

Jasanoff, Sheila (2005) *Designs on Nature*, Princeton, NJ: Princeton University Press.

Kasemir, B., Jäger, J., Jaeger, C. C. and Gardner, M. T. (eds) (2003) *Public Participation in Sustainability Science: A Handbook*, Cambridge: Cambridge University Press.

Lahsen, M. (2005) 'Technocracy, democracy and US climate politics: the need for demarcations', *Science, Technology and Human Values*, 30: 137–69.

McCright, A. M. and Dunlap, R. E. (2000) 'Challenging global warming as a social problem: an analysis of the conservative movement's counter-claims', *Social Problems*, 47: 499–522.

—— (2003) 'Defeating Kyoto: the conservative movement's impact on US climate change policy', *Social Problems*, 50: 348–73.

Mol, A. P. J. (2006) 'Environmental governance in the Information Age: the emergence of informational governance', *Environment and Planning*, C24: 497–514.

Pearce, F. (1991) *Green Warriors: The People and the Politics Behind the Environmental Revolution*, London: Bodley Head.

Priest, S. H. (2001) *A Grain of Truth: The Media, the Public, and Biotechnology*, Lanham, MD: Rowman & Littlefield.

Stirling, A. and Mayer, S. (1999) *Re-thinking Risk: A Pilot Multi-Criteria Mapping of a Genetically Modified Crop in Agricultural Systems in the UK*, Brighton: Science Policy Research Unit.

Winickoff, D., Jasanoff, S., Busch, L., Grove-White, R. and Wynne, B. (2005) 'Adjudicating the GM food wars: science, risk and democracy in world trade law', *Yale Journal of International Law*, 30: 81–123.

Yang, G. (2003) 'The co-evolution of the Internet and civil society in China', *Asian Survey*, 43: 405–22.

—— (2005) 'Environmental NGOs and institutional dynamics in China', *China Quarterly*, 181: 46–66.

Yearley, S. (1993) 'Standing in for nature: the practicalities of environmental organisations' use of science', in Milton, K. (ed.) *Environmentalism: The View from Anthropology*, London: Routledge, 59–72.

—— (1999) 'Computer models and the public's understanding of science: a case-study analysis', *Social Studies of Science*, 29: 845–66.

—— (2005a) *Cultures of Environmentalism*, Basingstoke: Palgrave Macmillan.

—— (2005b) *Making Sense of Science: Science Studies and Social Theory*, London: Sage.

12

Public participation and dialogue

Edna F. Einsiedel

According to William Gilbert, pioneering researcher into electricity and magnetism in the 17th century, a discussion with the public on science and technology was like listening to 'the maunderings of a babbling hag' (Anon. 2004). While this view may still be shared by a few today, by and large there has been a shift in the role and place of publics in discussions about science and technology. This chapter explores the landscape of public engagement and dialogue, and raises further questions about what these changes may mean for the place of science and technology in society.

Calls for public engagement

The 'participation explosion', as it has been called, has embraced a variety of concepts, illustrating the range of activities and categories with which such a term is associated, including 'citizen involvement', 'stakeholder engagement', 'participatory technology assessment', 'indigenous people's rights', 'local community consultation', 'NGO intervention', 'multi-stakeholder dialogue', 'access to information', or 'access to justice', all of which have been considered as cornerstones of sustainable development (Pring et al. 2002).

Numerous international documents have specified the need for engaging publics in decision-making, notably on matters concerning management of the environment. The Aarhus Convention points out that there is no alternative to participation and openness in policy-making if policy-makers are to meet the requirements of sustainable development (UN ECE 1998). Similarly, the Biosafety Protocol, adopted by the signatories to the Convention on Biological Diversity in 2000, has important provisions regarding public awareness and participation. Much earlier, Agenda 21 is notable for having devoted separate chapters to involving many different groups including women, children, youth, indigenous people, NGOs, local authorities, workers and trade unions, business and industry, scientists and technologists, and farmers (UN 1992). The Rio Declaration on Environment and Development, Principle 10, says about participation on environmental issues that:

> Such issues are best handled with the participation of all concerned citizens, at the relevant level. At the national level, each individual shall have appropriate access to information concerning the environment that is held by public authorities, including information on hazardous materials and activities in their communities, and the opportunity to participate in decision-making processes. States shall facilitate and encourage public awareness and participation by making information widely available. Effective access to judicial and administrative proceedings, including redress and remedy, shall be provided.
>
> (UN 1992)

Three elements noted above are basic to participation: access to information, participation in decision-making, and judicial redress when necessary.

Parliamentary documents such as the UK's *Open Channels* (Parliamentary Office of Science and Technology 2001) exemplify this policy interest at the national level. In this case, two reasons are given for supporting such activities: support for greater democracy or remaking democracy to be more authentic in its approaches (Giddens 1998), and making better decisions.

For the scientific community, its 'licence to practise' can no longer be assumed – the extension of this licence comes about through processes that include public engagement and dialogue. Complementing this notion is the idea of public engagement as a means of quality control in the more recent conceptualisation of innovation in the context of 'post-normal science'. In such a context, where science delivers its goods to society, where uncertainty and imprecision are inherent in the subjects of inquiry, and an increasingly recognised part of the scientific landscape, the reliability of established facts becomes less secure and the principle for practice then becomes quality:

> For post-normal science, quality becomes crucial, and quality refers to process as much as to product. . . . Lacking neat solutions and requiring support from all stakeholders, the quality of decision-making is absolutely crucial for the achievement of an effective product in the decision.
>
> (Ravetz 1999: 649)

In this case, the determination of quality can only be accomplished through a plurality of perspectives and commitments. Publics become part of the larger community that participates in this 'extended peer review', where those with interests in, or who are affected by, the products of science become part of the consortium of evaluators (Funtowicz and Ravetz 1994).

Which publics?

We do not take for granted the term publics, assuming these are fixed and known entities. We recognise that publics vary by time and place and issue, and that they are also 'imagined groups', to borrow from Anderson's (1991) notion of 'imagined communities'. They are analytical constructs as much as they are rhetorical inventions.

They are products of contexts: the same individual can assume different roles at different times (or at the same time), so the same individual may behave differently as member of the citizenry or the community of consumers, as 'users and non-users' of technology (Oudshoorn and Pinch 2002), or as member of some organised interest (Michael 1998). In other words, publics are a complex and heterogeneous set of actors and relations that arise from particular contexts (Irwin and Michael 2003).

Some public consultation activities have evoked images of 'the ordinary citizen', or the 'disinterested citizen', the latter to distinguish this individual presumably from those involved in organised interests, or those with predefined 'stakes', or 'stakeholder groups'. With the increasing dominance of science and technology in major policy issues and the labelling of the 21st century as the 'century of biology' (Venter and Cohen 2004), some discussion has focused on the 'scientific citizen' (Irwin 2001) or the 'biological citizen' (Petryna 2002). More recently, the Human Genome Project and its aftermath have seen the notion of 'genetic citizenship' elaborated (Heath et al. 2004). These constructions of citizenship revolve around questions of identity, but there are also the social networks that act to extend the different configurations in which members of publics operate.

There has been a growing literature on social movements as exemplified by environmental and patient organisations, sometimes called the third sector of knowledge production (Sclove 1995), with their distinct and growing capacities for influence on science and technology research and innovation trajectories. The 'civic epistemologies' that civil society organisations and individual citizens construct (Jasanoff 2005) are further building blocks in the knowledge society to which many publics contribute. These more nuanced ways of thinking about publics have important implications for public participation and dialogue.

The place of dialogue in public 'talk'

In thinking about the process of communications on science and technology, a frequently invoked spectrum is the one-way versus two-way flow of communication, information provision versus the dialogic approach. This dichotomy has been characterised as the deficit versus the interactive model of publics, where the former sees publics as empty vessels waiting to be filled with (scientific) wisdom; the second approach views publics as active participants in the science communication process. Each of these models suffers from the weight of its own caricature. Instead of a dichotomy of public participation, a continuum of participation is more appropriate. This continuum includes information provision, consultation (where publics are asked for feedback), involvement (where publics participate in the determination of appropriate solutions), and empowerment (where participation extends from defining the problem to determining the solution) (see e.g. Arnstein 1969). Just as we have emphasised publics as products of their contexts, the processes of communication, production, and use of knowledge develop out of context and circumstance. Publics are constant producers and receptors of communication and information; they imbibe, select, reject and develop expertise out of necessity or interest; they also remain unaware, blissfully ignorant or simply disinterested, by choice or happenstance.

While the dominant model in science communications remains linear, treating audiences as deficient in scientific knowledge, an important effort to acknowledge the different nuances behind public understandings and indeed, public expertise, has succeeded in moving the pendulum to the other side. A confluence of different factors contributed to this: recognition of the limits of scientific expertise, greater acknowledgement of uncertainty in science, a history of failed initiatives where public 'outrage' resulted from decisions that excluded those who were affected or who bore the risks; the growing influence of social movements that provided other sites of expertise; debates about the impacts of technologies, which led to questions about equity and the distribution of risks and benefits, control and participation in technological decisions. As a result, the questions of who participated, and how, led to different experiments in communication forms and the exploration of mechanisms for publics to talk back to science (see Chapter 5 in this volume).

Trends

In considering public dialogue on science and technology, we can identify four trends in terms of forms, frequency, timing and content.

Forms of engagement

Most striking about experiments with public dialogue are the different forms that exemplify such discussions with various publics. Such fora as consensus conferences (Joss and Durant 1995), citizen juries (Crosby 1995; Coote and Lenaghan 1997), scenario workshops (Andersen and Jaeger 1999) and deliberative mapping (Burgess et al. 2007) make use of different procedures, depending on the goals, the problem being addressed, or even the degree of scientific controversy. At the same time, all of these rest on dialogic processes, engagement with a knowledge base, discussion with a variety of experts, explicit recognition of values, and deliberation on and recommendations for preferred futures.

To illustrate, a *consensus conference* is one forum that allows citizens to be involved in the assessment of technology. The conference is a dialogue between experts and citizens and is open to the public and the media. The *citizen panel* plays the leading role: it consists of 15–25 people who are introduced to the topic by a professional facilitator. The citizen panel formulates the questions to be taken up at the conference, and participates in the selection of types of expert to answer them. The panel has two weekends for this preparation.

The selection of the *expert panel* is done in a way that facilitates airing of the essential and diverse range of views. The experts present their answers to the questions from the citizen panel, with further opportunities for clarification and more discussions between the expert panel, the citizen panel and the audience. The citizen panel spends the rest of the second and the third days working to produce a final document, with their conclusions and recommendations. Consensus on positions and recommendations, achieved through open discussion, is an important element of the process. On the fourth day, the citizen panel presents its report to the experts and the audience, including the media.

It is striking that many countries have employed this form of deliberation, more recently on gene technologies (e.g. Einsiedel et al. 2001; Seifert 2006; for a summary see The Loka Institute website, www.loka.org). We speculate that the public nature of such an approach allows for broader societal discussion potentially to occur around a controversial application (in this instance, GM food or cloning), and draws attention to what citizens view as key issues around an application that might be close to commercialisation, or is already in the marketplace. *Scenario workshops* start with a problem in search of solutions. These solutions may vary: they can be different technological approaches, regulatory solutions, or new ways of organising or managing problems. A set of scenarios is usually prepared before the workshop, which describes alternative ways of dealing with the problem (for example, different ways of disposing of nuclear waste). These scenarios offer starting points for discussion. Ideally, the scenarios presented should be reasonably differentiated according to the organisational or technical solutions presented, as well as their accompanying social and political values.

The four groups of actors involved usually include policy-makers, business representatives, experts and citizens. The participants discuss, critique and evaluate technological and non-technological solutions to the problems, and they can develop alternative visions of the future as well as proposals or action plans for realising them. Three phases are embedded within this process: a critical analysis phase, a visionary phase, and an implementation phase.

These approaches exemplify deliberative models but we recognise that the spectrum of public participation includes a broad continuum, ranging from more one-way approaches such as focus groups, to those involving more dialogue and discussion; they can utilise different technological approaches such as online discussions, teleconferencing and face-to-face fora.

Frequency

The frequency of deliberative engagement has been documented, perhaps in an unsystematic way. One can look at the number of public participation activities over the past two decades and see the increasing interest in such activities, with some suggesting this has become an industry in itself. We see this in the proliferation of the citizens' consensus conference approach in a variety of countries, on a range of gene technology issues, but most commonly on the controversial application of genetically modified food (Seifert 2006). On the subject of radioactive waste management, we also see a range of approaches to public participation with different dialogue tools, from more one-way approaches (public opinion surveys) to more deliberative forms (Kemp et al. 2006).

Timing

In terms of timing, public views and assessments of technology in the past had typically occurred at the back end, or at the point of commercialisation. In many instances this was 'too late' for technology developers or policy-makers, who had invested considerable effort and resources into development, only to find resistance from, or rejection by, the public. Today, it is not unusual for publics to be looking over the shoulders of scientists or technology developers at the earliest stages of research. The push to get publics involved much earlier in the technology development process, referred to as 'upstream engagement', exemplifies the shift in emphasis of the timing of such activities (Wilsdon and Willis 2004; Stilgoe et al. 2006). However, such upstream activities have to go beyond simple transfer of the same forms of analytical, deliberative processes used downstream; rather, this requires the engagement of publics in different sets of questions (Pidgeon and Rogers-Hayden 2007) and experimenting with a variety of engagement tools as required by context, issue and policy considerations.

Content

While the dialogue fora have increased and become more varied, we can also say something about their content. Many citizen panel reports springing from the use of consensus conferences on a variety of gene technology issues (most prominently on GM food) have raised questions about health impacts and safety only as starting points. However, other issues raised, and recommendations made, elaborate on notions of fairness, social justice, technological control, the global economy, environmental impacts and sustainability. These reports have clearly demonstrated the broader range of values that lay citizens bring to bear on technological questions.

Frames of understanding for public dialogue and deliberation

The work on public dialogue has been underpinned by theoretical frameworks that are not just descriptive or explanatory, they are also often normative.

Deliberative democracy

Work on deliberative democracy, also known as public realm theory, has been inspired by political theorists such as Hannah Arendt (1958) and Jürgen Habermas (1984). Both expressed concerns with modern-day democracies, but for different reasons. Arendt was concerned about the 'rise of the social' (which she equated with non-political and hence non-public elements). Habermas, on the other hand, was worried about the 'structural transformation' of contemporary societies, which he interpreted as the replacement of discursive and interactive politics of the past with present-day technical and administrative politics. An institutional arena of public discourse and civic participation is essential, they argue, to counterbalance the dual pressures of state and market. The public sphere is conceptualised both as a process of deliberation by the public, and as an arena or space in which this takes place. The normative element of their work has been geared to the creation of a robust public sphere, as well as preserving it from the erosive influences of modern society.

Habermas's concept of the public spere is authenticated through several requirements, including open access, voluntary participation, the development of public judgement through groups of citizens engaged in political deliberation, and the freedom to express opinions and to criticise the state. Other theorists, such as John Rawls (1971), have also pointed to a number of other conditions: adequate information, an absence of manipulation of processes or outcomes, an orientation toward solutions that address common as opposed to self-interests, and a condition of political equality in which 'the force of the argument' rather than power and authority is the norm.

Deliberation as the foundation for practical, collective decision-making processes has been cited as providing the roots for democratic legitimacy. Deliberative processes can impart new information, help individuals to order their preferences coherently, and allow reflexivity on individual preferences, enabling participants to adopt an 'enlarged mentality' (Benhabib 1998).

Critics of deliberative democracy, on the other hand, dismiss deliberative democracy as an impractical ideal that does not take into account practical challenges to the citizen: the costs of being informed (which may not be sufficient to counterbalance expected benefits), or the costs of engagement in terms of time and the scope of one's interests. Second, there may be a difficult-to-resolve tension between effectiveness and inclusiveness. The scale of many deliberative activities simply excludes a large number of citizens, while at the same time allowing an opportunity for a few to engage with an issue, be more deeply informed, and be able to debate and discuss an issue. Third, some claim that the norm of rationality and the force of the 'better' argument, alongside the disregard for interests, is simply unrealistic. Fraser (1992), for example, suggests the existence of contradictory public spheres where political power is not so easily 'bracketed and neutralized' (ibid. 115).

One can also take the same virtues applied to deliberative democracy – acquisition of new information, assistance for individuals in ordering their preferences coherently, and provision for reflexivity on individual preferences – and ask whether such results might be found among decision-makers. Has an 'enlarged mentality' in fact resulted in policy arenas? At present, we know little about whether this is an outcome.

Public participation as technology assessment

Other research on public engagement and dialogue on science and technology has typically focused on assessments of controversial technologies. As such, dialogue activities have been framed as one of the building blocks for doing constructive technology assessment (CTA) (Rip et al. 1995; Schot 2001). In a similar vein, participatory technology assessment (Joss and Bellucci 2002) puts the focus on the lay and stakeholder participants as key 'technology assessors'. While traditional technology assessment has focused on evaluating impacts and attempted to mitigate problematic outcomes, constructive technology assessment tries to influence the technology development process at the front end by focusing on shaping the technology's design. This pragmatic approach recognises that governments are typically engaged in promotion of technology as well as its regulation or control. These functions can be, and often are, incompatible when they are not considered in tandem, or are not integrated. For policy-makers, the practical question then becomes how to stimulate the development of technology while at the same time considering how to maximise its potential benefits and minimise its negative consequences.

Within the CTA framework, public dialogue is one of a range of activities designed to 'broaden the design of new technologies' through feedback processes into actual technology design and construction. Such feedback occurs through three activities: sociotechnical mapping, which combines stakeholder analysis with systematic examination and analysis of technical considerations; early and controlled experimentation, through which unanticipated impacts can be identified and potentially minimised; and dialogue between technology developers and publics to clarify public expectations and concerns, or 'the demand side of technology development'. To evaluate and shape CTA processes, three criteria are offered: anticipation, reflexivity and social learning. Not surprisingly, CTA was developed by policy communities (specifically the Netherlands Office of Technology Assessment) as an anticipatory approach to technology management. However, CTA does not go beyond technology design to ask the more difficult question for many publics: why this technology and not an alternative?

A complementary approach to CTA is offered through the perspective of the social shaping of technology. Scholars in this area are united by an insistence that the 'black box' of technology must be opened, to allow the socio-economic patterns embedded in both the content of technologies and the processes of innovation to be exposed and analysed (MacKenzie and Wajcman 1985; Bijker and Law 1992).

The social-shaping framework draws from a number of perspectives that share analytical concerns and more critical perspectives. This framework examines the ways in which social, institutional, economic and cultural factors shape the direction, as well as the rate of innovation, the form of technology (its content, practices and assumptions), and its outcomes. Through the varieties of public deliberation, dialogue and debates, such questions and challenges have sometimes forced designers or policy communities to refashion technological design (e.g. Oudshoorn and Pinch 2002).

Changing views of expertise and knowledge production

Public dialogue has also been viewed through the lenses of the deconstruction of expertise and the social distribution of knowledge(s). This body of work has tried to redefine expertise away from the narrow domain of the technical and to locate it in broader societal lacunae (e.g. Irwin and Michael 2003; Nowotny 2003). This is what Fischer (2003) called society's 'socio-cultural rationality', challenging the narrow technical rationality of expert frames. The authority of expertise has also been labelled as a form of 'delegated authority' that requires being held to norms of transparency and deliberative accountability (Jasanoff 2003). What Jasanoff (2005) has labelled 'civic epistemologies' is embodied differently across different cultures, using such criteria as styles of knowledge-making, accountability and the foundations for demonstrating expertise.

Institutional contexts

Finally, an understanding of the institutional contexts provides another framework within which to examine public engagement and dialogue. Drawing from institutional theory, institutions can be viewed as systems of rules and norms that can act to facilitate or constrain public participation. In this context, one can ask such questions as: Who has access and under what conditions? How are agendas or problems defined? What rules and norms operate to enhance or constrain deliberation? How are these reflected in procedural practices? Collectively, these systems of tacit and explicit ways of doing things can shape what Tarrow (1994) has called 'opportunity structures' that influence public dialogue norms and practices.

As part of institutionalising public dialogue and participation practices, one would necessarily have to describe shared goals, which may be achieved through these practices. In their analysis of over 200 public participation cases on environmental issues, Beierle and Cayford (2002) suggested that these goals, used as standards for assessing the impact of public participation, include the following: the incorporation of public values into decisions; the improvement of the substantive quality of decisions; conflict resolution among competing interests; building trust in institutions; and educating and informing publics. Their analysis found that these goals have contributed to the improvement of environmental policy; have diminished conflict and minimised public mistrust, which had characterised many environmental issues; and have contributed to better problem-solving processes. Public motivation to participate and institutional responsiveness have been key factors in this success.

Public dialogue and policy

I began this chapter by pointing to the increasing interest in policy communities in engaging publics in communication and dialogue. Admittedly, some of these activities have sometimes been driven by the need for legitimation of preselected policy choices, and some outcomes have sometimes been greeted with discomfort or dismay, if not distrust. The expectation among decision-makers is that public engagement is a way of demonstrating they have 'done the right thing', that they have taken appropriate notice of public views, that they have hit the right policy notes. Brian Wynne (2006) has argued, however, that policy-makers are 'hitting the notes but missing the music'. Expanding on this idea, his colleagues describe the currency of engagement in these terms:

> Engagement is about opening up policy, exposing it to criticism, challenging its assumptions (including those about knowledge and expertise) and forcing governments to make difficult decisions out in the open.
>
> (Stilgoe et al. 2006)

One of the striking characteristics of citizen dialogue and citizen reports is the breadth they demonstrate in determining what a technology means. This points to the continuing gap between the policy world and its heavy reliance on science-based risk assessment metrics, and citizen views, which bring in notions of equity and access, ethics, control and sustainability. For policy-makers, this will remain a significant and continuing challenge – how to integrate these different social metrics with the continuing reliance on a narrower vision of technology assessment.

If we look at public participation activities as discursive arenas, we need not limit ourselves to citizen-publics; members of organised interests – exemplified by civil society or non-governmental organisations – are very much part and parcel of communities of publics. The deliberative dimensions or engagement of public dialogues are enacted as a confrontation of discourses played out within and across the public sphere's networks of civil society. Dialogue and debates of varying discursive strains compete simultaneously for policy attention and initiative. Viewed in this way, everyone can play an important role in the deliberative processes that lead to policy-making without going through requirements of formally structured and controlled events. This might be regarded as deliberative democracy in more horizontal and socially distributed form.

This broader perspective also responds to criticisms of deliberative democracy theory as requiring participants to leave their interests outside the public arena, so to speak, in the spirit of reasoned exchange. It is precisely these preformed interests that encourage people to enter the public sphere at all. Requiring that they disengage themselves from their interests is unrealistic and unfair. The representation of both 'disinterested' and interested individuals might be better ensured through engagement and dialogue in its various forms, represented through various groups and individuals, and their different actions and activities.

Similarly, if we view policy communities as discursive arenas, the question of impacts then shifts from trying to understand 'whether and how deliberations by citizens influence policy' to 'how can the discourses of policy be reframed through the process of deliberation?' (Barnes et al. 2004). The pessimistic view is that we have a longer way to go. The optimist says we are at least on the way.

Suggested further reading

Burgess, J. and Chilvers, J. (2006) 'Upping the ante: a conceptual framework for designing and evaluating participatory technology assessments', *Science and Public Policy*, 33: 713–28.

Maasen, S. and Weingart, P. (2005) *Democratization of Expertise? Exploring Novel Forms of Scientific Advice in Political Decision-Making*, Dordrecht: Springer.

Pidgeon, N. and Rogers-Hayden, T. (2007) 'Opening up nanotechnology dialogue with publics: risk communication or upstream engagement?' *Health, Risk and Society*, 9: 191–210.

Wilsdon, J. and Willis, R. (2004) *See-through Science: Why Public Engagement Needs to Move Upstream*, London: Demos.

Other references

Andersen, I.-E. and Jaeger, B. (1999) 'Scenario workshops and consensus conferences: towards more democratic decision-making', *Science and Public Policy*, 26: 331-40.

Anderson, B. (1991) *Imagined Communities: Reflections on the Origin and Spread of Nationalism*, London: Verso.

Anon. (2004) 'Going public' (Editorial), *Nature*, 431: 883.

Arendt, H. (1958) *The Human Condition*, Chicago, IL: University of Chicago Press.

Arnstein, S. R. (1969) 'A ladder of citizen participation', *Journal of the American Planning Association*, 35: 216–24.

Barnes, M., Newman, J. and Sullivan, H. (2004) 'Power, participation, and political renewal: theoretical perspectives on public participation under New Labour in Britain', *Social Politics*, 11: 267–79.

Beierle, T. and Cayford, J. (2002) *Democracy in Practice: Public Participation in Environmental Decisions*, Washington, DC: Resources for the Future Press.

Benhabib, S. (1998) 'Toward a deliberative model of democratic legitimacy', in Benhabib, S. (ed.) *Democracy and Difference: Contesting the Boundaries of the Political*, Princeton, NJ: Princeton University Press.

Bijker, W. and Law, J. (eds) (1992) *Shaping Technology/Building Society: Studies in Socio-Technical Change*, Cambridge, MA: MIT Press.

Burgess, J. and Chilvers, J. (2006) 'Upping the ante: a conceptual framework for designing and evaluating participatory technology assessments', *Science and Public Policy*, 33: 713–28.

Burgess, J., Stirling, A., Clark, J., Davies, G., Eames, M., Staley, K. and Williamson, S. (2007) 'Deliberative mapping: a novel analytic-deliberative methodology to support contested science policy decisions', *Public Understanding of Science*, 16: 299–322.

Cartagena Protocol on Biosafety (2000) *Convention on Biological Diversity*, UN Environmental Program; www.cbd.int/biosafety/protocol.shtml

Coote, A. and Lenaghan, J. (1997) *Citizen Juries: Theory into Practice*, London: Institute for Public Policy Research.

Crosby, N. (1995) 'Citizens' juries: one solution for difficult environmental questions', in Renn, O., Webler, T. and Wiedemann, P. (eds) *Fairness and Competence in Citizen Participation: Evaluating Models for Environmental Discourse*, Dordrecht: Kluwer Academic.

Einsiedel, E., Jelsøe, E. and Breck, T. (2001) 'Publics at the technology table: the consensus conference in Denmark, Canada and Australia', *Public Understanding of Science*, 10: 1–6.

Fischer, F. (2003) *Reframing Public Policy: Discursive Politics and Deliberative Practices*, Oxford: Oxford University Press.

Fraser, N. (1992) 'Rethinking the public sphere: a contribution to the critique of actually existing democracy', in Calhoun, C. (ed.) *Habermas and the Public Sphere*, Cambridge, MA: MIT Press.

Funtowicz, S. and Ravetz, J. (1994) 'The worth of a songbird: ecological economics as a post-normal science', *Ecological Economics*, 10: 197–207.

Giddens, A. (1998) *The Third Way: The Renewal of Social Democracy*, Cambridge: Polity Press.

Habermas, J. (1984) *The Theory of Communicative Action, Vol. 1: Reason and the Rationalization of Society*, Boston, MA: Beacon Press.

Heath, D., Rapp, R. and Taussig, K. (2004) 'Genetic citizenship', in Nugent, D. and Vincent, J. (eds) *Companion to the Handbook of Political Anthropology*, London: Blackwell, 152–67.

Irwin, A. (2001) 'Constructing the scientific citizen: science and democracy in the biosciences', *Public Understanding of Science*, 10: 1–18.

Irwin, A. and Michael, M. (2003) *Science, Social Theory, and Public Knowledge*, Maidenhead: Open University Press.

Jasanoff, S. (2003) '(No) accounting for expertise', *Science and Public Policy*, 30: 157–62.

—— (2005) *Designs on Nature: Science and Democracy in Europe and the United States*, Princeton, NJ: Princeton University Press.

Joss, S. and Bellucci, S. (2002) *Participatory Technology Assessment: European Perspectives*, London: Centre for the Study of Democracy, University of Westminster in association with TA Swiss.

Joss, S. and Durant, J. (1995) *Public Participation in Science: The Role of Consensus Conferences in Europe*, London: Science Museum.

Kemp, R. V., Bennett, D. G. and White, M. J. (2006) 'Recent trends and developments in dialogue on radioactive waste management: experiences from the UK', *Environment International*, 32: 1021–32.

Maasen, S. and Weingart, P. (2005) *Democratization of Expertise? Exploring Novel Forms of Scientific Advice in Political Decision-Making*, Dordrecht: Springer.

MacKenzie, D. and Wajcman, J. (eds) (1985) *The Social Shaping of Technology: How the Refrigerator Got Its Hum*, Milton Keynes: Open University Press.

Michael, M. (1998) 'Between citizen and consumer: multiplying the meanings of the public understandings of science', *Public Understanding of Science*, 7: 313–27.

Nowotny, H. (2003) 'Democratizing expertise and socially robust knowledge', *Science and Public Policy*, 30: 151–6.

Oudshoorn, N. and Pinch, T. (2002) 'How users and non-users matter', in Oudshoorn, N. and Pinch, T. (eds) *How Users Matter: The Co-Construction of Users of Technology*, Cambridge, MA: MIT Press.

Parliamentary Office of Science and Technology (2001) '*Open Channels: Public Dialogue in Science and Technology*', Report 153, London: House of Commons.

Petryna, A. (2002) *Life Exposed: Biological Citizens after Chernobyl*, Princeton, NJ: Princeton University Press.

Pidgeon, N. and Rogers-Hayden, T. (2007) 'Opening up nanotechnology dialogue with publics: risk communication or upstream engagement?' *Health, Risk and Society*, 9: 191–210.

Pring, G., Zillman, D. and Lucas, A. (2002) 'The law of public participation in global energy and resources development', *Journal of Energy and Natural Resources Law*, 20: 79–83.

Ravetz, J. (1999) 'What is post-normal science?' *Futures*, 31: 647–53.

Rawls, J. (1971) *A Theory of Justice*, Cambridge, MA: Harvard University Press.

Rip, A., Misa, T. and Schot, J. (1995) *Managing Technology in Society: The Approach of Constructive Technology Assessment*, London: Pinter.

Schot, J. (2001) 'Towards new forms of participatory technology development', *Technology Analysis & Strategic Management*, 13: 39–52.

Sclove, R. (1995) *Democracy and Technology*, New York: Guilford Press.

Seifert, F. (2006) 'Local steps in an international career: a Danish-style consensus conference in Austria', *Public Understanding of Science*, 15: 73–88.

Stilgoe, J., Irwin, A. and Jones, K. (2006) *The Received Wisdom: Opening Up Expert Advice*, London: Demos.

Tarrow, S. (1994) *Power in Movement: Social Movements, Collective Action, and Mass Politics in the Modern State*, Cambridge: Cambridge University Press.

UN (1992) *Rio Declaration on Environment and Development (Agenda 21)*, 23.1-32.14, UN Department of Economic and Social Affairs; http://www.un.org/esa/sustdev/documents/agenda21/english/agenda21toc.htm

UN ECE (1998) *Convention on Access to Information, Public Participation in Decision-making, and Access to Justice on Environmental Matters*, Aarhus, Denmark: UN Economic Comission for Europe; www.unece.org/env/pp/documents/cep43e.pdf

Venter, C. and Cohen, D. (2004) 'The century of biology', *New Perspectives Quarterly*, 21: 73–7.

Wilsdon, J. and Willis, R. (2004) *See-through Science: Why Public Engagement Needs to Move Upstream*, London: Demos.

Wynne, B. (2006) 'Public engagement or dialogue as a means of restoring public trust in science? Hitting the notes but missing the music', *Community Genetics*, 9: 211–20.

13

Internet

Turning science communication inside-out?

Brian Trench

In the four decades since two university computers were first linked to each other over the prototype internet, scientific researchers have been innovators, early adopters and prolific adapters of internet technologies. Electronic mail, file transfer protocol, telnet, Gopher and the World Wide Web were all developed and applied first in research communities. The Web's development for sharing of information in the high-energy physics community unexpectedly heralded the internet's extension into many aspects of commerce, community, entertainment and governance. But despite the rapid proliferation and diversification of both over the past 15 years, the internet in its various forms has scientific communication indelibly inscribed into its fabric, and internet communication is thoroughly integrated into the practice of science.

This chapter reviews some effects of the internet's emergence as a principal means of professional scientific communication, and of public communication of science and technology. It notes several paradoxes that characterise these developments, for example the contradictory trends towards easier collaboration across continents, and towards greater fragmentation. It notes the very significant disturbances caused by electronic publishing in the all-important field of scientific journals. It suggests that these and other developments have made more completely porous than before the boundaries between professional and public communication, facilitating public access to previously private spaces, and thus 'turning science communication inside-out'.

At home in the internet

Already, a decade ago, it could be claimed that 'it is now difficult for scientists to remember how they worked without the internet' (Rowland 1998). Scientists are socialised into a world in which communication via the internet is 'natural'. Communication is the engine of science, accounting for an increasing amount of scientists' time and increasingly taking place over the internet. From posting calls for research proposals on the Web, through conferring with partners on a proposal by email, to the joint production, online submission and online review of the proposal via email and attachments, and on to the confirmation of the decision on the proposal, research projects or programmes can be – and are – established over the internet, without a face-to-face meeting or any paper changing hands. Very many of the routine activities of scientists are facilitated over the internet: calls for papers, editing of journals, hosting of conferences, sharing of data, authoring of papers, publication of conference proceedings and journals, and many more informal exchanges and encounters.

The processes by which the internet has come to fill this central place in internal scientific communication demonstrate well how technological innovations can be shaped socially to forms and functions not anticipated by the originators and first adopters. These processes have also been paradoxical in many ways: the internet facilitates collaboration between researchers on an effectively global scale across cultural, geographical and disciplinary boundaries; the internet also brings with it accelerated specialisation or 'balkanisation' within the sciences (Van Alstyne and Brynjolfsson 1996), as a sub-specialist who may be one of a kind in her own face-to-face community can interact with someone else in the same sub-specialism in any other part of the world. The internet operates as a means of collaboration, but also to facilitate and foster intensified competition, between institutions presenting their achievements over the Web, between discipline niches networking by email, and between individuals arguing in online discussion groups.

More important, in the context of public communication of science and technology, are the impacts – many of them also somewhat paradoxical – of the development of internet-based media on the dissemination of scientific information beyond the research communities. These media include versions of science news services already provided via print and broadcast, and 'net-native' services with their origins and their only manifestation in the internet environment. They cover promotional activities by research and educational institutions aimed at policy-makers and commercial partners, and public education initiatives by charities and science museums. Alongside research reports of new findings are other reports contesting or confirming those findings. Professional societies, research funders, higher education institutions, commercial companies, groups promoting science, groups challenging science, and many other interests are all active in amplifying or questioning information about science over the internet.

In enumerating these actors, we glide from internal scientific communication to public science communication, and it is here that we observe the most significant, and again most paradoxical, effects of the internet on science communication – principally the accelerated erosion of boundaries between previously distinct spheres of communication. While some science communication scholars have drawn attention to the long-time interpenetration of public and professional spheres (Lewenstein 1995; Bucchi 1998) and to the role of popularisation *within* science (Gregory and Miller 1998), these overlaps have become more tangible through the proliferation of Web-based science communication media.

'Access to the Web has opened up many aspects of scientific research previously hidden from the general public,' it has been claimed (Peterson 2001). Members of interested, but non-specialist, publics have access to information prepared by professional organisations primarily for consumption by professionals. Parts or all of sites maintained by scholarly societies and scientific journals are open-access, or require only that users register by name, opening internal discussions, pre-publication thoughts and professionals' agendas to wider public view. If, in some sense, the presentation of scientific information by scientists to the general public may be regarded as a performance, we can say – recalling sociologist Erving Goffman's terms – that, through the internet, some of the back-stage preparation has become visible to the prospective spectators of the front-stage performance (Goffman 1959).

Open access: for whom?

One means by which this happens is open-access publishing of journals and papers. This is one of the most far-reaching developments in professional scientific communication based on the internet. It is also one with significant implications for public communication of science and technology. Open-access publishing over the internet predates the development of the Web; one of the best known initiatives in this field, the Los Alamos Physics Papers (www.arxiv.org), was established in 1991. Despite several generations of technological change and interface design on the internet since then, the site retains its original look and feel. Here, papers are published electronically before formal review, or in parallel with their publication in a conventional journal. Any Web-user can access scholarly papers on particle accelerators, solar energy or cosmic rays – although, of course, not every Web-user can make sense of them.

This initiative, and similar ones, may be seen as an application of the 'communalist' norm that Merton (1942) identified over 60 years ago, although with enduring impact, as operating in scientific communities. The Human Genome Project's commitment to publish chromosome sequences as they were completed represented an application of that principle. Indeed, that project was impracticable without the internet; it was a computer and telecommunications software project as well as one in biological sciences. Sharing new information among scientists as widely as possible has found an appropriate, cost-effective platform and a means of effective realisation. The International HapMap Project, a successor to the Human Genome Project, started in 2002 to 'compare the genetic sequences of different individuals to identify chromosomal regions where genetic variants are shared'. In 2004, it removed its licensing regime to prevent potential commercialisation of its work: 'All of the consortium's data are now completely available to the public' over the internet (International HapMap Project 2004).

But the application of internet technologies has not been exclusively in the direction of communalism. The field of scholarly publishing in general, and scientific publishing in particular, has been in ferment for over a decade, as the small number of large-scale commercial journal publishers, on the one hand, seek to extract more value from their franchises through electronic publishing and database services, and professional societies, higher education institutions and their libraries, on the other hand, seek to apply the newer technologies to reducing the costs of sharing scientific information. That battle is still joined, but we can at least say that there is, as yet, no confirmation of Winston's 'law of the suppression of radical potential' of new communications technologies (Winston 1998).[1]

The internet's impact in this arena is keenly felt. The Berlin Declaration on Open Access to Knowledge in the Sciences and Humanities, initiated in 2003, begins:

> The internet has fundamentally changed the practical and economic realities of distributing scientific knowledge and cultural heritage. For the first time ever, the internet now offers the chance to constitute a global and interactive representation of human knowledge, including cultural heritage and the guarantee of worldwide access.
>
> (Berlin Declaration 2003)

The Declaration aims 'to promote the internet as a functional instrument for a global scientific knowledge base'. By early 2008, the directors or presidents of 250 research centres, universities and other institutions in 30 countries, mainly in Europe, had signed the declaration, and the collection of signatures was continuing. And this is just one of several such high-level international statements making a formal commitment to foster sharing of knowledge through internet-based publishing.

Scientific publishing revolt

These and other initiatives to transform the scientific publishing system and provide wider scholarly and public access to scientific information have been characterised as a 'revolt' (Meek 2001). The development of the internet put the inequity of the established scientific publishing system into sharp relief: 'Trawling computer databases [made] it possible for scientists to discover groundbreaking links between different research results which would previously have taken years of trawling through a jungle of indexes. The prospect of this incredible new tool being controlled by large private corporations has jerked scientists into action' (Meek 2001). Academic and research libraries have been to the fore in these actions, not least because of the large and continuing increases in journal subscription costs. Cornell University in the USA estimated that, between 1986 and 2001, the library budget at its main campus in Ithaca, New York, increased by 149 per cent; the number of periodicals purchased went up five per cent (Anon. 2004). The *New York Times* observed sympathetically: 'Some subscriptions cost thousands of dollars per year, and those journals are usually available online only to subscribers. This looks less like dissemination than restriction, especially if it is measured against the potential access offered by the internet' (Anon. 2003).

The Scholarly Publishing and Academic Resources Coalition (SPARC) has the backing of universities, research libraries and scholarly societies in seeking to cut journal subscription prices and to make publishers more responsive to customer needs. SPARC is associated with the publication of hundreds of journals in a wide variety of disciplines. One of the earliest was *Evolutionary Ecology Research*, whose editors moved the journal from a commercial publisher and relaunched it online, in collaboration with SPARC. The first editorial of the new-look journal declared that it wanted to 'maximize the number of scientists, scholars and students who have access to our articles ... [the journal] will be extraordinarily liberal in its dissemination rules' (Rosenzweig 1999). The application of these liberal rules means that much of the journal's content is available to those other than 'scientists, scholars and students', and this journal operates in a field that is contested both within science and within broader publics.

Research and academic libraries have also driven the development of the institutional repository model of networked publishing of papers generated from within their communities; these papers are now shared in a manner that attenuates the journals' control of the material generated from within those institutions. In the late 1990s, the advocacy group Public Library of Science asked researchers to commit to contribute only to journals that agree to place all their material into the public domain within six months of publication, securing tens of thousands of signatures worldwide. By the mid-2000s, the impetus had moved to scientific institutions and funders. From October 2005, it became a condition for researchers funded by the Wellcome Trust, a major funder of biomedical science in Britain, that they posted their papers on the life sciences archive PubMed Central within six months of publication.

Public Library of Science (PLoS) has gone on to launch its own journals: *PLoS Medicine* and *PLoS Biology* are published, open-access, on the internet. Similarly, BioMed Central publishes over 90 peer-reviewed journals online and free of charge; the publisher derives the income from charges to authors. Even Britain's Royal Society, the world's oldest scientific society, has added an online *Biology Letters* to its centuries-old *Proceedings*. Open-access journals represent a very small percentage of all scientific publishing, but the trend has continuing momentum.

Internet users, whether scientists, scholars, students or others, can find such resources through direct access to publishers' sites, or through portal services such as Stanford University's High Wire, Lund University's Directory of Open Access Journals, and PubMed Central, an open archive of literature from the biomedical and life sciences. Thanks to these and related developments, internet search engines include examples such as Google Scholar, which produces results from scientific journals and other forms of scholarly publication. Not merely are such search engines available without restriction to internet users, they are among the most intensively and extensively used facilities of the Web.

Opening formal scientific publication to public view brings with it some interesting challenges, both for scientific communities and for interested publics. For scientists, a key question, keenly contested, is how and whether the traditional standard of peer review should be applied in this changed environment. For Web users, a closely related question is that of the validation and interpretation of information found by hazard or by purposeful searching. It could be claimed, in a previous period, that peer review was the touchstone of scientific validity, and a critical boundary marker or gateway control in a 'continuity model' (Bucchi 1998; Chapter 5 in this volume) of science communication that envisages the arenas of communication as broadening progressively from intradisciplinary, to interdisciplinary, and beyond, to the general public. But the assumed gold standard of peer review has been tarnished by evidence of its misuse to protect entrenched positions and institutions, and by its failure to detect fraud. For Sense About Science, an advocacy group in Britain dedicated to 'an evidence-based approach to scientific issues', peer review remains 'an essential dividing line for judging what is scientific and what is speculation and opinion' (Sense About Science 2005). But even peer review's most ardent defenders have to acknowledge that it is a system under pressure, due to the proliferation of publishing outlets and the pressures on researchers to maintain certain levels of publishing productivity.

The unique status of peer review is also qualified by the development of internet-based outlets. Some scholars have argued for translation to the internet of the traditional peer review process (Harnad 1996), while others have presented a case for a radically modified 'cyberspace model', in which refereed and non-refereed services coexist (Giles 1996), or a 'scholars' forum' as an alternative publishing system (Buck et al. 1999). In 2003, the Royal Society established a working group to study best practice in communicating the results of new scientific research to the public. Among the questions the working group was asked to examine was 'what, if any, quality checks or filters should researchers subject their results to before communicating them to the public?' The very question indicates an acceptance that, in a changed communication environment, scientists cannot ensure that all scientific information reaching the public has been internally validated. The working group's deliberations were inconclusive: 'allowing reports of research results to be posted before they have been subjected to the full independent peer review process ... has clearly developed for the benefit of the researchers, [but] little consideration appears to have been given to the consequences of this practice for the public ... the potential for great damage clearly exists' (Royal Society 2006). In similar tone, the working group proposed on the operation of peer review: 'Further debate within the research community about the benefits and disadvantages of referee anonymity is desirable' (Royal Society 2006).

In 2005, three major medical journals – *The Lancet*, *Annals of Internal Medicine* and *BMJ* – opened their operations of peer review to the scrutiny of independent researchers, who examined the processing of over 1000 papers; early results represented a 'qualified thumbs-up to current editorial practices' (Giles 2006). As with the debate on modes of scientific publishing, the debate on the operation of peer review in the changed environment continues to be played out. In a development indicating the impact of the 'revolt' against commercial controls, a new initiative, Partnership for Research Integrity in Science and Medicine (PRISM) was started in the summer of 2007, with the support of the Association of American Publishers, to oppose government interventions favouring open-access publishing, specifically a proposed measure that would require peer-reviewed publications based on government-funded research to be 'surrendered' to government. The claimed aim of this coalition is to 'safeguard peer review', which it describes as 'a global standard for more than 400 years' (www.prismcoalition.org).

News for some, or for all?

Electronic publishing has partially opened the spectators' view to the backstage – partially, because only small proportions even of the 'interested publics' (Miller 1988) can use the information made available in this way. But the Web-publishing practices of very many scientific institutions have made access and use even easier. These institutions have adopted a public communication model, that of journalism, in the distribution of information. 'News', or some close equivalent, is a standard feature on websites generally, and many scientific institutions have adopted a journalism style of presentation to disseminate information about new developments, even where their primary purpose seems to be providing information from professional sources to professional audiences. Increasingly, research centres, scientific societies, research funders, universities and other higher education institutions directly employ science communicators or science writers to provide accessible summaries of research findings and other achievements, mainly via the internet. In this way, the institutions pre-empt, to some extent, the interpretive role of journalists working in the independent media, in a (partial) process of 'disintermediation'. Sharon Dunwoody (Chapter 2 in this volume) draws attention to the effects on science journalism of the 'shift to the internet'. I have written elsewhere (Trench 2007) on the ways in which this shift affects the status and role of journalists reporting science.

The internalisation by institutions of journalism forms can be seen in the websites of, for example, the European Commission's R&D services, the Institute of Physics, the Wellcome Trust and the Royal Society (in Britain), the Max Planck Society (Germany), Centre National de Recherche Scientifique (France) and the National Institutes of Health and the American Physical Society (USA): all have News, Research News, Actualités, Updates or News and Features directly at their home page, or easily accessible from that page. In some cases, the sites offer news alerts or similar services, often drawing on, or linking to, news reports in the general media. These institutions are professional societies, research bodies or research-funding agencies, thus with researchers as their prime public; in the cases of the French and German sites, they offer English-language versions as well as those in their own language, indicating that the global research communities are among their prime public. But these institutions' practices on the Web express further how the boundaries between professional and public communication are eroded, and provide a striking demonstration of a phenomenon that the German social theorist Jurgen Habermas observed over three decades ago in a very different information environment – that scientific communities use general news media to communicate with each other (Habermas 1971).[2]

Some of the sites maintained by scientific institutions or funders also present the source material on which the news item is based, or link their news reports to the relevant journal papers. Even where the links are not made directly, the more experienced Web-user can trace news media reports back to their proximate sources (press releases) and more remote sources (journal papers), thus laying bare the interpretation and reinterpretation in the processes of public dissemination.

However, the openness shown here is not found in what we might call less mature scientific cultures. In Ireland, a colleague and I surveyed over 100 websites published by scientific institutions, and found they used almost exclusively a one-way model of communication, rarely offering internet users the means to contribute to information and argument (Trench and Delaney 2004). Confirming the findings of comparable surveys in Germany and Poland (Lederbogen and Trebbe 2003; Jaskowska

2004), our study showed that scientific institutions used the Web much more to promote themselves to professional and business audiences than to share information about their activities with diverse social groups. Facilities for feedback or forum-type discussions were found on just three sites and, of these, one ceased to exist soon after the survey was completed. Applying criteria that reflect widely shared conceptions of internet best practice, we noted that just a quarter identified a particular person for contact purposes, and only one-tenth responded to a message sent to them; half the websites gave contact details for scientists, but just three named individual authors of individual pieces; three-quarters of the websites had a News section, but less than one-third of websites overall dated the posting of their content; three-quarters of the sites had Links sections, and most links indicated were to institutions of the same type as the originating site. Our assessment was that the publishers of these science sites were using the Web mainly to promote themselves to peers, partners and clients and, very much less, to communicate with diverse publics.

Scientific cultures and their associated communication practices may vary widely between institutions and countries, and these findings are not capable of being generalised to broader international arenas. Institutional policies are more-or-less restrictive on their individual members and more-or-less responsive to public contributions. The internet facilitates personal communication as well as formal, institutional communication, and this too is extensively represented in science through the use of newer internet technologies such as weblogs (blogs) to present individual views and facilitate open discussion. The internet facilitates multimedial, affective communication as well as text- or numbers-based dissemination of technical information, and this is represented, for example, in the use of podcasts on some science sites. The leading international scientific journals, *Nature* and *Science*, deploy both these facilities on their websites, although the blogs by writers attached to the journals generally take the form of diary notes or informal observations from assignments at major conferences rather than open-ended contributions to discussion, as blogging elsewhere generally tends to be. These websites are the shop-windows of publications and other services offered for sale; they use podcasts as a form of 'trailer' for material covered more formally in scientific papers or news reports in the pages of the journals. Elsewhere, as on the website of Imperial College London, podcasts present a package of interviews on current and recent research projects at that institution; production of podcasts is also a practical training exercise for the college's students of science communication.

Individual scientists' blogs are found most readily in thematic areas where there is significant public attention and debate, for example space, climate, energy, behaviour, ecology and genetic and reproductive technologies. The blogs may be seen as efforts by such individuals to go beyond the more detached institutional stance towards public concerns in these areas, perhaps representing a strongly held personal point of view. Strong opinions are generally the motivation also for science-watchers among the general public to maintain blogs or other fora of interactive communication on science and technology. Indeed, the promise of provocative content is sometimes the lure proposed by services that host, or provide gateways to, blogs.

A journalist with *Nature*, writing about her own reading of blogs for possible news stories, offers a confirmation of the central argument of this chapter: 'Blogs are windows into academic coffee room chatter of the sort the media is not normally privy to' (Tomlin 2007).

Making sense of science news

The continuing proliferation and diversification of internet communication on science and technology, the strategic, even political character of some scientific institutions' internet communication, and the self-serving and partisan character of some individual scientists' internet communication all accentuate a difficulty for internet users that has already been touched on: the very widely varying types of sources of available information present significant challenges to internet users in validating and interpreting such information. Some research has shown the increasing reliance of diverse publics, notably in the USA, on the internet as a source of information on science, medicine and technology (Lacroix 2001; Miller 2001). Personal observation of students indicates that, for a clear majority of science-interested people within certain age groups, the internet is a first (and often last) resource for information on current science-related topics. Yet internet searches on more-or-less any such topic will produce many types of document from a wide range of sources: the variety includes reviewed journal papers, but also self-published research reports, statements by interest groups, news media articles, company promotions, and contributions to mailing lists or internet news groups. It takes above-average internet literacy to distinguish these different types of information and informant from each other.

The challenge to internet users is intensified by the prominent presence on the Web of science advocates in, and on the borders of, the scientific communities. In Britain, groups such as the Institute of Ideas, the Science Media Centre, Sense about Science, and the Social Issues Research Centre use their websites to promote what they see as a correct scientific approach to current issues. For some such groups, their Web presence is a response to perceived counter-science tendencies in the mass media and in society as whole. However, other internet-based services, equally accessible to seekers of scientific information, lift the curtain more on science's backstage and invite public scepticism about the motives of much contemporary science. The Union of Concerned Scientists (UCS) and the Center for Science in the Public Interest (CSPI), both based in North America, take strong positions on the social responsibilities of science; UCS focuses on environmental issues, and CSPI on food safety and nutrition. Long-established groups such as the Bulletin of Atomic Scientists and the Pugwash Conference, both of them groups of scientists and other scholars concerned with nuclear threats to global security, have broadened their audiences through the internet. In presenting their case, all these groups offer a critique of practices within science and policies for science.

In the contested area of climate science, Real Climate provides a forum for scientists, but their discussions on new evidence and its interpretation offer public insights into previously obscured science-in-process. However, in space science, where there is unusually strong participation from amateurs, and public discussions are often intense, early experience of news groups and Web discussions persuaded astronomers to limit that view back-stage. Open disagreement in 1998 on the risk of asteroid 1997 XF11 colliding with Earth led to controversy within the space science community, not just about how such assessments are made, but also about the posting of asteroid impact predictions on the Web. The confusing controversy apparently brought no relief from the risks associated with premature judgements. In November 2000, the International Astronomical Union (IAU) carried on its website a statement that there was a one-in-500 chance that asteroid 2000 SG344 might collide with Earth on 21 September 2030. A NASA source underlined that the level of probability of a collision in this case was much higher than in any previous one. Despite the 'international expert review' of this assessment, it took just two days for 'additional information' to appear on the IAU website. This revealed that additional observations of 2000 SG344 from image archives indicated that 'the closest the object can approach the Earth in 2030 is 11 lunar distances on 23 September'. Since I gathered this material from the IAU website in 2000 (Trench 2003), the Union has become much more economical with its information. There are no longer such announcements, or any trace of previous announcements, on the IAU website. The page on near-Earth objects outlines IAU policy about the handling of reported sightings. Much more of the website's content is defined as members-only. News and published press releases relate mainly to organisational matters such as election of officers (www.iau.org). However, much of the material relating to those earlier events could be found in archives of Usenet (News) groups stretching back to the mid-1990s. Closing again the once-open curtain can never be fully effective.

In considering how interested publics deal with so much, and such diverse, scientific information on the internet, the case of biomedical science is especially sensitive. For here, the contents of professional communication can have 'end-user' value as diagnosis or remedy. Medline, the database of medical-scientific materials that is a primary research resource for medical professionals, can be accessed in somewhat reduced form free-of-charge. One early study of users of this resource showed that 30 per cent of users were not researchers, teachers or doctors, but others searching for medical information (Lacroix 2001). The National Library of Medicine established Medline-Plus to cater for these non-professional users by organising access to the database materials around topics in which users had shown repeated interest. Another US site, WebMD.com, presents 'trustworthy' information on a range of medical topics, as seen by experts in medicine, journalism and health communication; the site publishers state: 'We pride ourselves in knowing our audience's needs'. In Britain, the National Health Service publishes NHS Direct, with 'answers to your common health questions' (www.NHSDirect.org.uk). Even more obviously user-driven is a service such as IrishHealth.com, which offers users several ways to shape the content of the site.

Alongside information from such sources, which are at least partially based within the medical profession, internet users may find advice and remedies from drugs companies, advocacy and awareness groups, complementary medicine practitioners and mystics. Medical professionals fear that the easy availability of such information causes confusion and may encourage self-diagnosis. A Canadian medical information researcher worried that 'through the internet, patients not only have access to as much information as clinicians, but they are also starting to provide advice to other patients through websites that they host and manage and email lists that they browse freely' (Jadad 1999). A British researcher notes that publishers of personal homepages on health matters have a 'clear intention ... to be a provider, as well as a consumer of health information'. He suggests that the encouragement to Web-users to purchase treatment or advice represents a 'clear break with the Parsonian models of the doctor/patient relationship' (Hardey 2000).[3]

There are several initiatives in the biomedical publishing sector, such as the Hi-Ethics Consortium, and in the government sector, such as the Science Panel on Interactive Communication and Health, to establish standards for websites that would allow users to discern professional, and therefore credible, sources (McLellan 2000). This form of voluntary regulation seems most appropriate to the structures and cultures of the internet, but these initiatives have not been notably successful. They present, in any case, an obvious difficulty of monitoring or enforcement: a site conforming to agreed standards on one day, or on one document, may not do so on another.

Unavoidable uncertainty

Much of the science – notably the biological and medical science – that comes into the public domain is uncertain and contested, both from within and from outside science. These uncertainties and contests can no longer be hidden from public view. They may even become a main element of public sense-making of science news, as Dunwoody (1999) suggests:

> As increasing access to information on the Web and other outlets allows individuals to range beyond their hometown media, readers will be able to assemble meaning on a grander scale by cobbling together stories about the same topic from a variety of places. It seems almost inevitable that such triangulation will make uncertainty a common take-home message.

The Web's characteristics as a publishing medium, most obviously its juxtaposition of information and perspectives through search results and hyperlinking, may contribute to heightening the sense of uncertainty. But the Web also – and this is just another of its many paradoxes – provides platforms that are especially suited to the open, public negotiation of those uncertainties. By using the Web's hyperlinking capacities, and conscientiously connecting and comparing a range of perspectives and source-types, creative Web publishers can offer users a fuller picture and an understanding of the bases of uncertainty. They can do so, for example, by providing pointers to sources other than their own that may confirm or qualify the information for which they are directly responsible. They can assist the user further by adding information about information, indicating how their own information has been compiled, and offering responsible and critical assessment of competing claims and diverse contributions.

Other contributors to this volume (e.g. Irwin, Chapter 14) and to discussion of science communication more generally (e.g. Zehr 1999) have considered that uncertainty is a given, unavoidable condition of science in public. Whereas a traditional model of science communication supposed that uncertainty was progressively reduced, even eliminated, in the movement from relatively closed to relatively open spheres of communication, a contemporary model acknowledges both that much of the science of most interest to the public is inherently uncertain, and that science-interested publics – and perhaps even the citizenry in general – are capable and willing to handle such uncertainty. That was already the view of some citizen organisations in the context of the crisis over BSE ('mad cow disease') in Britain. On the day of publication of the official report into the handling of the crisis, the director of the National Consumer Council, Anna Bradley wrote:

> The chief problem [about the UK Government's stance on BSE] was that public statements appeared to be founded on certainty – that beef was safe to eat – but they were accompanied by an acknowledgement of uncertainty ... Uncertainty and ignorance are part of the normal spectrum of scientific data.
>
> (Bradley 2000)

Such a statement implies a role for public communication of science and technology that is more challenging and more complex than the popularising, much less evangelising, roles with which we are familiar. It is a role for which the internet provides both back-stage settings and front-stage areas of performance that meet the audience's needs and expectations.

Notes

1 Winston's law refers to the 'social constraints [that] coalesce to limit the potential of the device radically to disrupt pre-existing social formations'.
2 Habermas characterised scientific communities as 'bureaucratically encapsulated'; in these circumstances, general media provided a means of linking communities.
3 The reference is to the Canadian sociologist Talcott Parsons who, from the 1930s onwards, developed an analysis of the role of professionals that was especially influential in medical sociology (Parsons 1954).

Selected further reading

Friedman, S. M., Dunwoody, S. and Rogers, C. L. (eds) *Communicating Uncertainty – Media Coverage of New and Controversial Science*, Mahwah, NJ: Lawrence Erlbaum Associates. (See, in particular, contributions of Dunwoody and Zehr.)

Lewenstein, B. (1995) 'From fax to facts: communication in the cold fusion saga', *Social Studies of Science*, 25: 403–36.

Science Communication, 22, 3 (March 2001) Thematic Issue: 'Internet Bounty – how the public harvests science and health information'. (See, in particular, contributions of Lacroix, Miller and Peterson.)

Other references

Anon. (2003) 'Open access to scientific research' (Editorial), *New York Times*, 7 August.

Anon. (2004) 'Access all areas', *Economist*, 7 August.

Berlin Declaration (2003) *The Berlin Declaration on Open Access to Knowledge in the Sciences and Humanities*, www.zim.mpg.de/openaccess-berlin/berlindeclaration.html.

Bradley, A. (2000) 'Respecting our fears', *The Guardian*, 26 October. www.guardian.co.uk/analysis/story/0,387993,00.html.

Bucchi, M. (1998) *Science and the Media – Alternative Routes to Science Communication*, London: Routledge.

Buck, A. M., Flagan, R. C. and Coles, B. (1999) *Scholars' Forum: A New Model for Scholarly Communication*, http://library.caltech.edu/publications/scholarsforum.

Dunwoody, S. (1999) 'Scientists, journalists and the meaning of uncertainty', in Friedman, S. M., Dunwoody, S. and Rogers, C. L. (eds) *Communicating Uncertainty – Media Coverage of New and Controversial Science*, Mahwah, NJ: Lawrence Erlbaum Associates, 59–80.

Friedman, S. M., Dunwoody, S. and Rogers, C. L. (eds) *Communicating Uncertainty – Media Coverage of New and Controversial Science*, Mahwah, NJ: Lawrence Erlbaum Associates.

Giles, J. (2006) 'Journals submit to scrutiny of their peer-review process', *Nature*, 439: 252.

Giles, M. (1996) 'From Gutenberg to gigabytes – scholarly communication in the age of cyberspace', *Journal of Politics*, 58: 613–26.

Goffman, E. (1959) *The Presentation of Self in Everyday Life*, Garden City, NY: Doubleday.

Gregory, J. and Miller, S. (1998) *Science in Public*, New York: Plenum Trade Press.

Habermas, J. (1971) 'The scientization of politics and public opinion', in Habermas, J. (ed.) *Toward a Rational Society – Student Protest, Science and Politics*, London: Heinemann, 62–80.

Hardey, M. (2000) 'The "home page" and the challenge to medicine', *He@lth Information on the Internet*, 16 (August), 1–2.

Harnad, S. (1996) 'Implementing peer review on the net', in Peek, R. and Newby, G. (eds) *Scholarly Publication: The Electronic Frontier*, Cambridge, MA: MIT Press.

International HapMap Project (2004) *International HapMap Consortium Widens Data Access*, www.genome.gov/12514423

Jadad, A. (1999) 'Promoting partnerships: challenges for the internet age', *British Medical Journal*, 319: 761–4. http://bmj.bmjjournals.com/cgi/content/full/319/7212/761

Jaskowska, M. (2004) 'Science, society and internet in Poland', in *Scientific Knowledge and Cultural Diversity – Proceedings of the Eighth International Conference on Public Communication of Science and Technology*, Barcelona, June 2004, Barcelona: Rubes Editorial, 263–7.

Lacroix, E-M. (2001) 'How consumers are gathering information from MedlinePlus', *Science Communication*, 22: 283–91.

Lederbogen, U. and Trebbe, J. (2003) 'Promoting science on the web: public relations for scientific organizations – results of a content analysis', *Science Communication*, 24: 333–52.

Lewenstein, B. (1995) 'From fax to facts: communication in the cold fusion saga', *Social Studies of Science*, 25: 403–36.

McLellan, F. (2000) 'Ethics in cyberspace – the challenges of electronic publishing', in Jones, A. and McLellan, F., *Ethical Issues in Biomedical Publication*, Baltimore, MD: Johns Hopkins University Press.

Meek, J. (2001) 'Science world in revolt at power of journal owners', *The Guardian*, 26 May. www.guardian.co.uk/uk_news/story/0,3604,496855,00.html

Merton, R. (1942) 'The normative structure of science', reprinted in Merton, R. (1973) *The Sociology of Science: Theoretical and Empirical Investigations*, Chicago, IL: Chicago University Press.

Miller, J. (1988) 'Reaching the attentive and interested publics for science', in Friedman, S. M., Dunwoody, S. and Rogers, C. L. (eds) *Scientists and Journalists – Reporting Science as News*, Washington, DC: American Association for the Advancement of Science, 55–70.

—— (2001) 'Who is using the web for science and health information?', *Science Communication*, 22: 256–73.

Parsons, T. (1954) 'The professions and social structure', in Parsons, T. (ed.) *Essays in Sociological Theory – Pure and Applied*, Glencoe, IL: Free Press.

Peterson, I. (2001) 'Touring the scientific web', *Science Communication*, 22: 246–55.

Rosenzweig, M. (1999) Editorial, *Evolutionary Ecology Research*, 1, 1.
Rowland, F. (1998) 'Scientists in communication', in Scanlon, E., Hill, R. and Junker, K. (eds) *Communicating Science – Vol. 1: Professional Contexts*, London: Routledge, 55–60.
Royal Society (2006) *Science and the Public Interest – Communicating the Results of New Research to the Public*, London: Royal Society. www.royalsoc.ac.uk/displaypagedoc.asp?id=23615
Sense About Science (2005) *I Don't Know What To Believe – Making Sense of Science Stories*; London: Sense about Science. www.senseaboutscience.org.uk/index.php/site/project/29.
Tomlin, S. (2007) 'Blogging science', *Science & Public Affairs*, September: 23.
Trench, B. (2003) 'Les nouvelles scientifiques sur le Web: exploration de nouveaux espaces d'information', in Schiele, B. and Jantzen, R. (eds) *Les Territoires de la Culture Scientifique*, Lyon: Presses Universitaires de Lyon; Montréal: Les Presses de l'Université de Montréal.
—— (2007) 'How the internet changed science journalism' in Bauer, M. and Bucchi, M. (eds) *Journalism, Science and Society: Science Communication between News and Public Relations*, London: Routledge.
Trench, B. and Delaney, N. (2004) 'Public education on science: how Irish scientific institutions use the Web', paper presented at *Science and Mathematics Education for the New Century*, Dublin, 23–24 September 2004.
Van Alstyne, M. and Brynjolfsson, E. (1996) 'Could the internet balkanize science?', *Science*, 274, 1479–80.
Winston, B. (1998) *Media Technology and Society – A History from the Telegraph to the Internet*, London: Routledge.
Zehr, H. (1999) in Friedman, S. M., Dunwoody, S. and Rogers, C. L. (eds) *Communicating Uncertainty – Media Coverage of New and Controversial Science*, Mahwah, NJ: Lawrence Erlbaum Associates, 3–21.

14

Risk, science and public communication

Third-order thinking about scientific culture

Alan Irwin

Introduction

This chapter explores some different ways of thinking about science communication and risk management. In certain contexts, there has been a transition from 'first-order' (or deficit) models of science–public relations to a greater emphasis on public engagement and dialogue – discussed here as 'second-order' thinking (see Chapters 5 and 12 in this volume).

This chapter especially addresses certain problematic and challenging aspects of this partial movement between first- and second-order approaches. 'Third-order' thinking about risk, science and public communication asks fundamental questions about the underlying relationship between first- and second-order approaches, the changes that have taken place (both in theory and practice), and the future direction of scientific governance and science communication.

It is important to emphasise that first-, second- and third-order thinking are not presented here as distinct historical stages, nor as an inevitable sequence. This is not a story of one way of thinking inevitably giving way to the next, and then the next. Instead, the situation in most national and local contexts is of these different orders being mixed up (or churned) together. The deficit model coexists with talk of dialogue and engagement. While some organisations and individuals look for quick and easy solutions to communication problems, others have begun to reflect on the inherent limitations, contextualities and conditionalities of both deficit and dialogue.

Importantly, not all parties will agree on any particular categorisation: what one party might view as 'engagement' can often be seen as top-down communication by another (especially if disappointed with the outcome). Thus social experiments in 'public engagement' very often lead to accusations that the exercise was too restricted, too short and insufficiently democratic. From the perspective of this chapter, such accusations do not invalidate initiatives but can represent an essential resource within the public scrutiny of sociotechnical change.

This chapter presents the public communication of science and technology as more than a matter of communication style. Instead, through the device of third-order thinking, we confront basic issues of the shaping and direction of sociotechnical change, the frameworks within which communication takes place, cultures of governance and control (especially relating to the institutions of science) and the choices that are available to citizens within modern democracies.

Background: shifting keywords of science governance and science communication

Within western Europe in particular, something interesting has happened to the language of science communication and scientific governance (Hagendijk et al. 2005). In the UK, a landmark 2000 report from the House of Lords tackled the broad topic of 'Science and Society' by emphasising the 'new mood for dialogue' over science and technology (House of Lords 2000). Along with other British reports from the late 1990s onwards, the Lords Select Committee presented science's relationship with society as being under strain. Their Lordships called for a greater acknowledgement of doubt and uncertainty, and for a change in the culture of science communication and decision-making 'so that it becomes normal to bring science and the public into dialogue about new developments at an early stage' (ibid. 13). Meanwhile, the Netherlands has a longer history of dialogue and engagement around science and technology, but more recently has held a major public debate around genetically modified foods (Hagendijk and Irwin 2006). In 2002, the European Commission published its own action plan on science and society, and called for an 'open dialogue' over technological innovation as part of its 'new partnership' between science and society (European Commission 2002).

Throughout this recent period, one particular example of public dialogue and engagement has maintained (and indeed extended) its iconic status: the operation of Danish consensus conferences. As Anders Blok has noted (citing Rowe and Frewer 2005), in discussions of what has come to be known as 'deliberative democracy' (Dryzek 2000), consensus conferences 'normally sit alongside citizens' juries, negotiated rule-making, action planning workshops and citizen-based advisory committees, all examples of true public participation as opposed to mere communication mechanisms' (Blok 2007: 166). Consensus conferences have now been held in some 16 nations, all inspired by the notion that groups (generally 14–16) of citizens can bring important perspectives to bear on apparently technical issues (e.g. stem cell research, genetically modified foods, environmental problems). Following Blok, consensus conferences can be seen as promising civic virtues ('more informed, active, cooperative and thus "better" citizens'; ibid. 167); governance virtues (by enhancing the legitimacy of decisions); and cognitive (or epistemic) virtues (by articulating different perspectives, clarifying controversial areas, bringing 'citizen science' to problems; Irwin 1995).

These are, of course, ideals, and as such may not always be achieved. However, the basic inspiration for all these developments is the notion that more active, open and democratic relations between science and citizens are both desirable and necessary. At the same time, they suggest a critique of what has gone before, of what the UK

Chief Scientific Adviser describes in the Lords report as a 'rather backward-looking vision' (House of Lords 2000: 25) where 'difficulties in the relationship between science and society are due entirely to ignorance on the part of the public' and 'with enough public-understanding activity, the public can be brought to greater knowledge, whereupon all will be well.' This conceptualisation of an ignorant and uninformed public for science, christened the 'deficit theory' by social scientists in the 1990s (Wynne 1995; Irwin and Wynne 1996) has been a powerful provocation to change, prompting the argument that we now need to move 'from deficit to dialogue' (Irwin 2006a). In what follows, we present this conceptual and institutional shift as a transition (albeit a partial one) from first- to second-order thinking.

Quite how much things have changed within scientific governance and science–citizen relations can be gauged by a brief excursion back to the early 1990s, when Britain was edging towards what would later be seen as the BSE crisis. Certainly, the notion that science–citizen relations were badly mishandled by the relevant government department – and a consequent desire to 'avoid another BSE' – represent a powerful and continuing influence on institutional thinking about risk communication and management.

Mad cows and first-order thinking

When, in 1990, the (then) UK Ministry of Agriculture, Fisheries and Food (MAFF) and its government minister found themselves confronting a new category of risk, they responded in what was then conventional fashion. The issue was the risk to consumers from British beef: could eating contaminated meat lead to a human form of BSE (bovine spongiform encephalopathy)? Reassurances were offered by a variety of governmental and industrial groups, all attempting to convey the message that the risks were minimal and that consumers could purchase British beef with confidence. Famously, the British and international media featured the minister feeding a beefburger to his daughter before a cluster of eager photographers. As one advertisement from the Meat and Livestock Commission put it:

> Eating British beef is completely safe. There is no evidence of any threat to human health caused by this animal health problem (BSE) . . . This is the view of independent British and European scientists and not just the meat industry. This view has been endorsed by the Department of Health.
>
> (*The Times*, 18 May 1990)

Writing just a few years later, I picked out a number of major characteristics of this first-order exercise in risk communication (Irwin 1995: 53). First of all, we can identify an authority claim based on the language of certainty. Second, science is presented as absolutely central to the whole issue. The use of 'independent scientists' in the above quotation appeals to an apparent faith that science can be trusted in a manner that would not necessarily apply to the meat industry (or government). Third, these efforts at risk communication do not draw on public engagement in any meaningful way. Consumers are to be protected, rather than consulted. This is 'top-down' (or 'one-way') communication. Fourth, this 'science-centred' approach to risk management and risk communication takes little account of the diversity, nor the possible knowledgeability, of publics. This final point became especially apparent when subsequent control measures were brought into play without any meaningful consultation with the abattoir workers responsible for putting them into practice (and who might have pointed out that the operating conditions of the abattoir are rather different from those of the carefully controlled laboratory).

To these analytical points, we should add the not-insignificant observation that this mode of risk communication was conspicuously unsuccessful. Meat sales suffered and governmental credibility was damaged. Official claims of certainty were substantially undermined by very public disagreements among scientists concerning the scale of risk. When, subsequently, human cases of variant CJD (Creutzfeldt–Jakob disease, linked to BSE) began to appear, previous expressions of official confidence were judged to be inappropriate and irresponsible. All this was summarised by the inquiry into BSE and variant CJD, which published its report a decade later:

> The Government did not lie to the public about BSE. It believed that the risks posed by BSE to humans were remote. The Government was pre-occupied with preventing an alarmist over-reaction to BSE because it believed that the risk was remote. It is now clear that this campaign of reassurance was a mistake. When on 20 March 1996 the Government announced that BSE had probably been transmitted to humans, the public felt that they had been betrayed. Confidence in government pronouncements about risk was a further casualty of BSE.
>
> (Phillips et al. 2000: Vol. 1, section 1)

Terms such as 'unwarranted reassurance' (ibid. 1150) and 'culture of secrecy' (ibid. 1258) recur throughout the Phillips report (which at one point goes so far as to suggest that the main object of MAFF's communication strategy was 'sedation' of the publics; ibid. 1179). Recurrent within the report also was the perceived need by civil servants and others to counteract what were anticipated to be 'alarmist' public and media reactions to the existence of risk. In a highly significant section, it is argued that 'the approach to communication of risk was shaped by a consuming fear of provoking an irrational public scare' (ibid. 1294). In contrast, the then Chief Scientific Adviser is quoted favourably as arguing that the temptation 'to hold the facts close' so that a 'simple message can be taken out into the market place' should be resisted. Instead, 'the full messy process whereby scientific understanding is arrived at with all its problems has to be spilled out into the open' (ibid. 1297). In this way a 'culture of trust' rather than one of secrecy can be developed. More broadly, the official report stressed several points that, in the wake of BSE, have become central to the language of scientific governance in many European countries, notably the Netherlands, Denmark and the UK (for a discussion across eight European nations, see Hagendijk et al. 2005; for a discussion of such issues in a global and 'developmental' context, see Leach et al. 2005):

- 'Trust can only be generated by openness';
- 'Openness requires recognition of uncertainty, where it exists';
- 'The public should be trusted to respond rationally to openness';

- 'Scientific investigation of risk should be open and transparent';
- 'The advice and reasoning of advisory committees should be made public'.

(Source: Phillips *et al.* 2000)

From first to second order?

The story of BSE in the UK is therefore also the story of a larger movement in thinking about risk management and risk communication. The public stance adopted by MAFF in 1990 represented an almost classical representation of first-order thinking about risk communication. By 2000, the official inquiry could draw on this case in order to advocate much greater transparency and openness – especially in terms of acknowledging uncertainty and respecting the rationality of the public, rather than fearing alarmism. Although not especially emphasised within the report itself, the abolition of MAFF and its replacement with a new government department, the Department for Environment, Food and Rural Affairs (Defra), also marked a greater willingness to engage in a two-way relationship with the wider publics: in other words, an emphasis on dialogue and deliberation rather than deficit. As was noted above, greater transparency, recognition of uncertainty and public engagement have been advocated by a series of British reports from the late 1990s onwards (RCEP 1998; DTI 2000; RS/RAE 2004; see also Irwin 2006a).

One central argument within this chapter is that the movement between first- and second-order thinking is (or should be) more than a matter of changing communication style. It is for this reason that I have employed here the language of different orders, as opposed to relying on the more common language of 'deficit' and 'dialogue'. Rather than simply replacing the language of deficit with that of dialogue, each approach (at least potentially) draws on deeper intellectual and political roots.

Looked at in more conceptual terms, first-order thinking depends strongly on what many social theorists (notably Bauman 1991; Giddens 1991; Beck 1992) have described as the culture of modernity, a culture within which science is presented as the embodiment of truth and the task of government becomes one of bringing rationality to human affairs. As Bauman and Beck have argued, notions of uncertainty and ambivalence can fit uneasily within such a culture, given its characteristic confidence in science-led progress. Indeed, Bauman has suggested that the whole substance of modern politics and modern life is the quest for order and, as he strongly puts it, the 'extermination of ambivalence'. Certainly, the first-order perspective accommodates well with the pragmatically expressed but technocratically derived view of government as being primarily concerned with bringing rational principles to bear on political and social challenges. At the same time, the conventional (positivistic) understanding that science can speak 'truth to power' (Jasanoff 1990, 2005) reinforces the idea that the wider public can, of epistemological necessity, play only a restricted role in deciding about risk issues.

Whether justified in terms of the requirement to make policy in difficult and changing circumstances, or as a fundamental belief in the objectivity and rationality of science, the first-order approach draws on well established intellectual and political principles. In addition, the economic significance of risk issues cannot be ignored, as was especially apparent during the BSE crisis of the 1990s, where the impact on British agriculture was clearly a significant governmental and industrial concern. The 'rational' case for economic development and the first-order case for science-led progress appear to have worked reasonably harmoniously with one another – at least until relatively recently (Ezrahi 1990).

What of the second-order approach to risk communication and management? The general argument here is that second-order thinking necessitates more than a softening of the first-order communication style or a new form of public relations campaigning (although for a critical perspective on second-order thinking, see Mouffe 1993, 2000).

In the first place, the move towards greater transparency and engagement can be connected to larger discussions about the merits of deliberative democracy and the need to revitalise political institutions – discussions associated with Jürgen Habermas, John Rawls and John Dryzek, among other theorists (Rawls 1972; Habermas 1978; Dryzek 2000; Hagendijk and Irwin 2006). Second, and linked to this point, Beck and his contemporaries have emphasised the manner in which conventional political institutions have come under great challenge when dealing with issues of risk. While modernistic institutions might present themselves as being 'in control', the evidence from public protests over genetically modified foods, nuclear power and road-building programmes suggests that this may not be the case (Beck 1992). In such circumstances, demands for new forms of democratic accountability and engagement become a central characteristic of contemporary political life. Third, the emergent emphasis on trust, transparency and two-way communication has been stimulated by a series of social scientific studies from the 1990s onwards, which suggest a more fundamental institutional challenge in dealing with contested areas of risk (Irwin and Wynne 1996). In contrast to the conventional portrayal of public groups as irrational and uninformed, various empirical studies have explored the knowledgeability and resourcefulness of particular publics when encountering science-related issues within the contexts of everyday life (Brown 1987; Epstein 1996; Kerr et al. 1998; Bloor 2000). One practical implication of such research has been that, rather than simply presenting wider society as an impediment to scientific and technological progress, it is necessary to examine critically the operational assumptions and practices of scientific institutions: the very form of reflexive scrutiny that first-order approaches generally evade. Seen from this perspective, second-order thinking may have become a practical necessity if public policy is to be made – and justified – in circumstances of social and technical uncertainty (Stilgoe et al. 2006).

Of course, and as Chantal Mouffe implies, the intellectual differences between first- and second-order thinking can be exaggerated. In principle, commitments to science and science-led progress, and to transparency and dialogue, do not necessarily contradict one another. After all, many areas of science and technology remain relatively uncontroversial and enjoy apparent public support. The main argument of this chapter, however, is that, equally, it cannot be simply assumed that they are fully commensurate with one another. Furthermore, while it might be tempting to suggest that the emphasis on trust, openness and engagement has somehow replaced older, outmoded forms of practice, I argue in the remainder of this chapter that a more complex (and confusing) situation now operates in which first- and second-order approaches operate in uneasy coexistence and unconsidered juxtaposition.

Rather than simply being a matter of shifting from one communication style to another (or a straightforward story of first-order approaches giving way to second-order), I want instead to suggest that the relationship between risk, science and public communication raises more profound questions of scientific and political culture. My argument is that these questions are generally neglected amidst the institutional enthusiasm for new risk communication and public engagement approaches. It is this larger discussion that lies at the heart of third-order thinking, to which we will return after the next section.

Putting second-order thinking into practice

One route into these issues is through consideration of the experience of putting second-order thinking into practice within institutional processes that have operated conventionally according to first-order principles (Irwin 2001). The 'GM Nation?' debate on the commercialisation of genetically modified crops in the UK can be used by way of brief illustration (for a comparison with the parallel Dutch debate, see Hagendijk and Irwin 2006; see also Chapter 17 in this volume). Taking place during the summer of 2003, the debate was designed to be 'innovative, effective and deliberative' but also 'framed by the public'. Its broad aim was to 'provide meaningful information to Government about the nature and spectrum of the public's views, particularly at grass roots level, to inform decision-making.' Briefly put, the eventual conclusions of the debate report were: people are generally uneasy about GM; the more people engage in GM issues, the harder their attitudes and the more intense their concerns; there is little support for early commercialisation; there is widespread mistrust of government and multinational companies. In summary, the report characterises public opinion over the commercialisation of GM as 'not yet – if ever' (Heller 2003: see also Understanding Risk Team 2004).

Here, then, we have what was, in terms of UK practice, the most developed application of second-order thinking within the policy-making process thus far. However, the evidence suggests that, rather than representing a shift from older ways of thinking to newer ones, the situation is best described as one of uneasy coexistence. As I consider it, this is less to do with specific aspects of the debate's design, although such matters can be significant in themselves, than with the wider political and institutional framework within which the debate was conducted (Irwin 2006a). Thus there is an apparent tendency for governmental institutions, in particular, to view second-order discussions as a discrete phase within the policy process: an activity to be fed into decision-making at the appropriate time, alongside other forms of evidence, before business as usual can return. Such an approach imposes fundamental constraints on second-order engagement with science and risk, and especially limits second-order perspectives within the frameworks previously established by first-order understandings. We can offer a few illustrations of these points:

> The 'public' strand of the debate ran in parallel with a separate review of the available science and an economic assessment of the costs and benefits of GM crops. It would appear that the construction of public debate, economic and

> scientific reviews as three separate strands inhibited the possibility of transparent public engagement in 'technical' analysis or of public discussion openly reflecting upon the issues raised by the other streams.
>
> (Irwin 2006b)

Equally, the UK Government's eventual decision to proceed with GM technology on an 'individual case-by-case basis' fitted more easily with the economics and science strands than with the very cautious public debate. While the economics and science strands were generally accepted and appeared to feed directly into the government decision, the 'public' strand was presented as a matter for further deliberation and debate – a viewpoint for government to bear in mind, rather than a body of firm evidence on which it must act.

It was also very apparent in the debate that members of the public typically 'framed' the underlying issues much more broadly than did governmental and industrial officials. While for the concerned civil servants this was a matter of deciding about a particular technical issue, for many members of the public, the debate was connected to a wider-ranging set of questions about the power of transnational companies, globalisation, the future of British agriculture, and the comparative benefits of innovation to North American industry and British consumers. While policy-makers tended to frame the issue as a matter of 'risk' (to humans and the environment), this by no means captured the full spectrum of public assessments, and may in itself represent a first-order framing of the underlying issues (Wynne 2002). When risk did figure as an issue in public debate, this was generally balanced against questions of 'need', rather than being presented as a rational calculus of relative harm (Jones 2004).

Important issues arise concerning the timing of such debates. Thus the organisation of discussion late in the process of technology development substantially impeded the possibilities for constructive engagement over the technology itself including, in this case, alternative approaches to food and farming (Wilsdon and Willis 2004). However, even the most ambitious 'upstream' debate is likely to be ineffective unless the more fundamental issues raised here are properly addressed.

Finally, any hope that such an exercise in engagement would lead to social consensus was certainly disappointed. One characteristic outcome of engagement exercises is that they lead to further accusations and arguments: in this case, suggesting both that the exercise was 'hijacked' by activist groups and that it was far too restricted in participation, depth and coverage (House of Commons, Environment, Food, and Rural Affairs Committee 2003). This was even more apparent in the Dutch national debate on the same issue, when activist groups simply withdrew their cooperation in protest at the manner in which the debate was being framed (Hagendijk and Irwin 2006). This also suggests that what one party (typically the institutions of government) might present as 'engagement' can be viewed by another as 'old-style' deficit thinking. Like beauty, 'engagement' can lie in the eye of the beholder. Of course, this is not necessarily a negative characteristic (or a condition that can be cured with just a little more attention to detail). Disagreement and controversy can also be seen as bringing energy, excitement and focused attention to debates – and in that sense can be viewed as an important societal resource.

From the perspective of more reflexive third-order thinking about risk, science and public communication, one clear implication of the 'GM Nation?' debate is that there are many aspects of the relationship between first- and second-order perspectives that remain unexplored and neglected. Certainly, for this example at least, we can dispense with the idea that there has been anything so straightforward as a paradigm shift from one approach to another. It would appear instead that, even in the case of the UK (where criticisms of the deficit theory have been taken very seriously), and even in an area such as GM policy (which is especially well developed in this regard), the shift to second-order thinking has been partial, fixed-term and patchy.

A similar conclusion applies to the iconic example of Danish consensus conferences mentioned at the start of this chapter. Despite the external representation of consensus conferences as one of the most developed examples of second-order thinking, within Denmark there has been an academic debate over their impact and effectiveness (and, among certain politicians, even their continued operation) (Horst 2003; Jensen 2005; Blok 2007). Once again, we are reminded that these changes in the language and practice of science communication/scientific governance do not represent an inevitable, once-and-for-all historical sequence, but rather a sometimes-contested focus for discussion, disagreement and societal reassessment.

Third-order reflections

Third-order thinking in this context does *not* refer to a new model of scientific governance or science communication that will resolve the problems created by first- and second-order perspectives. Instead, it represents a move away from sloganising about what is best, towards more critical reflection – and reflection-informed practice – about the relationship between technical change, institutional priorities and wider conceptions of social welfare and justice. It is very important to stress that this is not simply a matter of categorising individual activities and initiatives into one order or another. As we have suggested, what might appear as dialogue to one party can look remarkably like deficit to another. Equally, this is not about developing a new, improved toolkit (although tools can be very useful), but rather is about interrogating the operating assumptions and modes of thought on which individual initiatives depend, and considering the practical and conceptual implications of this. Third-order thinking also takes us away from the notion that any given approach to communication is necessarily and intrinsically superior. Instead, deciding what is appropriate to any particular situation must be a matter for contextual judgement, but also for recognition of the limitations and strengths of all approaches. Put succinctly, third-order thinking invites us to consider what is at stake within societal decisions over science and technology, and to build on the notion that different forms of expertise and understanding represent an important resource for change rather than an impediment or burden (Stilgoe et al. 2006; see Table 14.1).

Some third-order thoughts about the discussion so far will help to substantiate these general points. In the first place, it is apparent that the expressed institutional enthusiasm for second-order approaches has not been accompanied by systematic

Table 14.1 Characteristics of first-, second- and third-order thinking on risk communication

	First order	*Second order*	*Third order*
Main focus	Public ignorance and technical education	Dialogue, engagement, transparency, building trust	Direction, quality and need for sociotechnical change
Key issues	Communicating science, informing debate, getting the facts straight	Re-establishing public confidence, building consensus, encouraging debate, addressing uncertainty	Setting science and technology in wider cultural context, enhancing reflexivity and critical analysis
Communication style	One-way, top-down	Two-way, bottom-up	Multiple stakeholders, multiple frameworks
Model of scientific governance	Science-led, 'science' and 'politics' kept apart	Transparent, responsive to public opinion, accountable	Open to contested problem definitions, beyond government alone, addressing societal concerns and priorities
Sociotechnical challenge	Maintaining rationality, encouraging scientific progress and expert independence	Establishing broad societal consensus	Viewing heterogeneity, conditionality and disagreement as a societal resource
Overall perspective	Focusing on science	Focusing on communication and engagement	Focusing on scientific/ political cultures

and considered attention to the policy implementation of such approaches, nor to the challenges this would potentially generate. Instead, and as was evident within the UK GM debate, the tendency has been to view transparency and engagement as ends in themselves, and as a supplement to conventional procedures, rather than as a means of developing a more meaningful scrutiny of prevailing modes of scientific governance. Equally, the assumption that openness and engagement will restore institutional credibility (as opposed to revealing more fundamental antagonisms around social and technical change) remains unchallenged.

Second, and linked to that point, we are likely to witness growing criticism of policy approaches that adopt second-order rhetoric but without considering the fuller implications of such a perspective. Viewed sceptically, this seems to be an inevitable consequence of second-order thinking – engagement generates the demand for further engagement, and transparency leads to accusations of opacity (Horst 2003). In more immediate terms, the implication is that institutions should not promise more than they can deliver – which implies making explicit the limitations to openness and engagement as well as the constructive possibilities. This is especially important when the contemporary demand for transparency and dialogue must sit alongside the unavoidable requirement for accountability and leadership. A commitment to openness and democracy should not imply an abnegation of institutional responsibility, nor that complex issues of socio-scientific decision-making should always be turned over to a referendum. Instead, new forms of leadership are required that are open and transparent, but also capable of defending chosen courses of action in full acknowledgement of significant areas of uncertainty and the existence of alternative strategies and perspectives.

Third, the suggestion of this chapter has been that questions of communication strategy with regard to risk and science cannot be uncoupled from larger matters of scientific governance and scientific culture. The language of transparency, two-way exchange, trust and uncertainty can, it is true, be employed in entirely instrumental and superficial terms. Such an approach is likely to generate rather than (as it is intended) appease anxieties. Equally, claims of this sort can signal a temporary departure (or diversion) within the policy process. This may offer some progress from the old deficit theory, but ultimately will operate according to the same principles and assumptions – and at best will represent institutional 'listening' rather than dialogue. What such instrumental or diversionary approaches to risk, science and communication neglect are the wider questions of the direction, quality and need for socio-scientific change that engagement exercises so typically generate.

Third-order thinking about the character of scientific culture and the possibilities for sociotechnical change is not presented here as a panacea to public concerns around technical change, nor as a new policy mode. Instead, put at its simplest, it suggests that science–public relations need to be placed in wider context, and that critical evaluation is required of current approaches to scientific governance and science communication, whether of the first- or second-order variety, or any unstable compound thereof. As the case of BSE in Britain suggests, changes have taken place in the institutional treatment of science communication and risk issues in particular. The argument of this chapter is that critical thinking about the significance of such changes, and about the need for further change, has been rather slower in coming. Such thinking will require attention not simply to the mechanics of science–public relations, but also to deeper questions such as the relationship between scientific governance, political economy and innovation strategy, and the operation of national policy processes in an increasingly globalised setting. In recognising the partiality of progress from first- to second-order thinking, we also raise issues that take us to the core of social and scientific 'progress' in democratic societies.

To offer one final example of these issues in practice, nanotechnologies are being presented both as possessing huge potential for social benefit (for example, cancer-tackling nanobots and other medical possibilities) and for threat (as all human life is reduced to 'grey goo'). Perhaps the dominant political response has been to point to the huge potential of the nanotechnologies and to advocate increased public education in the issues (supported by social surveys suggesting that only a small proportion of the population even recognise the term 'nanotechnology'). However, second-order thinking has also found expression, and strong arguments have been made for democratic engagement and scrutiny of the nanotechnologies (RS/RAE 2004; Kearnes et al. 2006).

Looked at from a third-order perspective, it is hard to deny the benefits of education or the value of democratic discussion. However, what both these generally neglect is a deeper scrutiny of the possibilities for regional and national autonomy within the worldwide economy, the relationship between the nanotechnologies and societal values and preferences (in all their diversity), the strategies being adopted right now by international corporations, and the manner in which current processes of scientific governance serve to assist or hinder the expression of democratic principles. In this situation, the public communication of science and technology both takes on new significance and faces substantial new challenges. More importantly, perhaps, new possibilities emerge for forms of communication that do not simply trade in the unreflexive language of deficit and dialogue, but that open up fresh interconnections between public, scientific, institutional, political and ethical visions of change in all their heterogeneity, conditionality and disagreement.

Suggested Further Reading

Epstein, S. (1996) *Impure Science: AIDS, Activism and the Politics of Knowledge*, Berkeley, CA: University of California Press.

House of Lords, Select Committee on Science and Technology (2000) *Science and Society*, London: Stationery Office.

Irwin, A. (1995) *Citizen Science: A Study of People, Expertise and Sustainable Development*, London: Routledge.

Jasanoff, S. (2005) *Designs on Nature: Science and Democracy in Europe and the United States*, Princeton, NJ: Princeton University Press.

Leach, M., Scoones, I. and Wynne, B. (eds) (2005) *Science and Citizens: Globalization and the Challenge of Engagement*, London/New York: Zed Books.

Stilgoe, J., Irwin, A. and Jones, K. (2006) *The Received Wisdom: Opening Up Expert Advice*, London: Demos.

Other references

Bauman, Z. (1991) *Modernity and Ambivalence*, Cambridge: Polity Press.

Beck, U. (1992) *Risk Society: Towards a New Modernity*, London: Sage.

Blok, A. (2007) 'Experts on public trial: on democratizing expertise through a Danish consensus conference', *Public Understanding of Science*, 16: 163–82.

Bloor, M. (2000) 'The South Wales Miners Federation: miners' lung and the instrumental use of expertise, 1900–1950', *Social Studies of Science*, 30: 125–40.

Brown, P. (1987) 'Popular epidemiology: community response to toxic waste induced disease in Woburn Massachusetts', *Science, Technology and Human Values*, 12: 76–85.

DTI (2000) *Excellence and Opportunity: A Science and Innovation Policy for the 21st Century*, London: Department of Trade and Industry/Stationery Office.

Dryzek, J. S. (2000) *Deliberative Democracy and Beyond: Liberals, Critics, Contestations*, Oxford: Oxford University Press.

Epstein, S. (1996) *Impure Science: AIDS, Activism and the Politics of Knowledge*, Berkeley, CA: University of California Press.

European Commission (2002) *Science and Society Action Plan*, http://europa.eu.int/comm/research/science-society/pdf/ss_ap_en.pdf

Ezrahi, Y. (1990) *The Descent of Icarus: Science and the Transformation of Contemporary Democracy*, Cambridge, MA: Harvard University Press.

Giddens, A. (1991) *Modernity and Self-identity: Self and Society in the Late Modern Age*, Cambridge: Polity Press.

Habermas, J. (1978) *Towards a Rational Society*, London: Heinemann.

Hagendijk, R. and Irwin, A. (2006) 'Public deliberation and governance: engaging with science and technology in contemporary Europe', *Minerva*, 44: 167–84.

Hagendijk, R., Healey, P., Horst, M. and Irwin, A. (2005) *Report on the STAGE Project: Science, Technology and Governance in Europe*, STAGE, www.stage-research.net

Heller, R. (2003) *GM Nation? The Findings of the Public Debate*, London: Department of Trade and Industry.

Horst, M. (2003) 'Controversy and collectivity – articulations of social and natural order in mass mediated representations of biotechnology', PhD thesis, Copenhagen Business School, Doctoral School on Knowledge and Management, Department of Management, Politics and Philosophy.

House of Commons, Environment, Food and Rural Affairs Committee (2003) *Conduct of the GM Public Debate, 18th Report of Session 2002–2003*, London: Stationery Office.

House of Lords, Select Committee on Science and Technology (2000) *Science and Society*, London: Stationery Office.

Irwin, A. (1995) *Citizen Science: A Study of People, Expertise and Sustainable Development*, London: Routledge.

—— (2001) 'Constructing the scientific citizen: science and democracy in the biosciences', *Public Understanding of Science*, 10: 1–18.

—— (2006a) 'The politics of talk: coming to terms with the "new" scientific governance', *Social Studies of Science*, 36: 299–320.

—— (2006b) 'The global context for risk governance: national regulatory policy in an international framework', in Bennett, B. and Tomossy, G. (eds) *Globalization and Health: Challenges for Health Law and Bioethics*, Amsterdam: Springer, 71–85.

Irwin, A. and Wynne, B. (eds) (1996) *Misunderstanding Science? The Public Reconstruction of Science and Technology*, Cambridge: Cambridge University Press.

Jasanoff, S. (1990) *The Fifth Branch: Science Advisers as Policymakers*, Cambridge, MA: Harvard University Press.

—— (2005) *Designs on Nature: Science and Democracy in Europe and the United States*, Princeton, NJ: Princeton University Press.

Jensen, C. B. (2005) 'Citizen projects and consensus building at the Danish Board of Technology: on experiments in democracy', *Acta Sociologica*, 48: 221–35.

Jones, K. E. (2004) 'BSE and the Phillips report: a cautionary tale about the update of "risk", in Stehr, N. (ed.) *The Governance of Knowledge*, New Brunswick/London: Transaction Publishers: 161–86.

Kearnes, M., Macnaghten, P. and Wilsdon, J. (2006) *Governing at the Nanoscale: People, Policies and Emerging Technologies*, London: Demos.

Kerr, A., Cunningham-Burley, S. and Amos, A. (1998) 'The new human genetics: mobilizing lay expertise', *Public Understanding of Science*, 7: 41–60.

Leach, M., Scoones, I. and Wynne, B. (eds) (2005) *Science and Citizens: Globalization and the Challenge Of Engagement*, London/New York: Zed Books.

Mouffe, C. (1993) *The Return of the Political*, London/New York: Verso.

—— (2000) *The Democratic Paradox*, London/New York: Verso.

Phillips, Lord, Bridgeman, J. and Ferguson-Smith, M. (2000) *The BSE Inquiry: The Report*, London: Stationery Office, www.bseinquiry.gov.uk

Rawls, J. (1972) *A Theory of Justice*, Oxford: Clarendon Press.

Rowe, G. and Frewer, L. (2005) 'A typology of public engagement mechanisms', *Science, Technology and Human Values*, 30: 251–90.

RCEP (1998) *Setting Environmental Standards: 21st Report*, London: Royal Commission on Environmental Pollution/Stationery Office.

RS/RAE (2004) *Nanoscience and Nanotechnologies: Opportunities and Uncertainties*, RS policy document 19/04, London: Royal Society/Royal Academy of Engineering.

Stilgoe, J., Irwin, A. and Jones, K. (2006) *The Received Wisdom: Opening Up Expert Advice*, London: Demos.

Understanding Risk Team (2004) *An Independent Evaluation of the GM Nation? Public Debate about the Possible Commercialisation of Transgenic Crops in Britain, 2003*, Understanding Risk working paper 04-02, www.risks.org.uk

Wilsdon, J. and Willis, R. (2004) *See-through Science: Why Public Engagement Needs to Move Upstream*, London: Demos.

Wynne, B. (1995) 'Public understanding of science', in Jasanoff, S., Markle, G. E., Petersen, J. C. and Pinch, T. (eds) *Handbook of Science and Technology Studies*, Thousand Oaks, CA: Sage, 361–88.

—— (2002) 'Risk and environment as legitimatory discourses of technology: reflexivity inside out?', *Current Sociology*, 50: 459–77.

15

Public communication of science and technology in developing countries

Hester du Plessis

The academic writing that formally identified society's need to be informed about scientific findings, as well as scientists' need to explain science to society, originated within the developed world when science began to exert an overt influence on the daily life of people. The same need became apparent in developing countries over a similar period. The topic of science communication, together with debates around the development of science and technology and the need for the public to understand science, appeared in policy documents of colonial as well as post-colonial governments in India and Africa. Public communication of science and technology was seen as a means of both advancing and understanding science.

The well-documented historical environment inherited from the developed world supported and informed the initial desire to bridge the communication gap between the world of science and that of the general public. Scientists and academics from both worlds designed models through which the public was studied, science reporting was evaluated and science communication was facilitated (Hountondji 1983; Miller 1983, 1998; Durant 2000; Raza et al. 2002; Du Plessis and Raza 2004; Riana and Habib 2004).

Reflection on the scientist's responsibilities was fostered and supported by sociologists in India and philosophers in Africa, resulting in an acknowledgement that the process that falls between the gathering of scientific facts and the assimilation of these facts is much more complex than previously thought. A differentiated and multidisciplinary academic approach, reflecting the various strategies for science communication – 'practical scientific literacy', 'civic scientific literacy' and 'cultural scientific literacy' – became part of the communication process in developing worlds (Shen 1975).

The theme of science communication in developing regions has become one of the most prominent in the field of public communication of science, signalled by, among other things, the attention it receives in leading sources about research in developing areas, such as Scidev.net.

Initiatives, debates and – more recently – studies of science communication and public engagement with science have multiplied in many areas of the developing world, often in parallel with the growth of research and development (R&D) investments, with such variety and richness that it would be implausible to map them thoroughly within the space available here.

In this chapter, a short overview is presented on some of the main issues raised by the development of science and technology communication processes, using the cases of India and Africa and of South Africa in particular.

India

Attention to the importance of science education, science writing and dissemination of science findings in Indian society has a long history. Two chief parallel streams of activity can be identified historically: one focused on the protection and preservation of ongoing science practised on a daily basis by local communities (known as indigenous knowledge systems), the other concerned with the promotion and propagation of the 'big sciences'.

From 1761 to 1903, the British in India embarked on 'the great surveys', which championed the imperial programme of expropriation and control (Cohn 1996; Riana and Habib 2004). 'Popular science' as a theme was promoted since the early 19th century by several leading public figures, including the Nobel Laureate poet Tagore (1861–1941), who published his Bengali *Visyaparichay* in 1937. The universities in India were instituted from 1857 onwards, and as early as 1876 the Indian Association for the Cultivation of Science was initiated. In 1914, the Indian Science Congress and the National Institute of Science (currently known as the National Science Academy) were founded. Ramchandra, who was active in the mid-19th century, was a pioneer of Urdu science journalism during the colonial period. Raja Rammohun Roy (Sarkar 1975) promoted mass education and introduced modern science in schools.

Riana and Habib (2004) described factors that shaped science in late-colonial India and the considerable contributions from a multidisciplinary group of scholars who studied a 'social history of science' and who undertook the problematic study of the politics of scientific knowledge in India. These works included studies on the cultural appropriation of modern science in India and on the modalities of the dialogue between different systems of scientific knowledge. With the departure from a dominant positivist conception of science and the new conceptual possibilities opened up by notions such as Kuhn's paradigm (Kuhn 1962) and other insights into the philosophy of science, Indian scholars 'acquired a more nuanced perspective on the processes of critical assimilation, cultural redefinition and reinvention of science' (Riana and Habib 2004: viii).

Since Indian independence in 1947, the attention given to science and the progressive growth of science communication has been well documented. Soon after independence a government-driven process of social transformation was initiated by Jawaharlal Nehru (the so-called Nehruvian era). He introduced the notion of fostering a 'scientific temper' among the Indian population.[1] According to Nehru, the essence of scientific temper is an active, sensitive, questioning understanding and a creative relationship between man and his environment. It is a rational approach to the discovery of truth through free and creative thinking, experimentation and objective analysis, and a steadfast commitment (with humility, not arrogance) to established truth. At the same time, it recognises the tentative and continuously unfolding character of scientific understanding of phenomena, disentangling the different forces and motivations at work (Pattnaik 1992). A consistency is required between theory and practice, which excludes the practice, also condemned by Hountondji (1983), of vague qualitative thinking.

Two types of science communication have been identified in India during this period. One type included institutional (governmental) science communication initiatives and science education programmes, such as those conducted under the auspices of the National Council for Science and Technology Communication (NCSTC), the National Centre for Science Communicators and the National Council of Science Museums. The second type continued the tradition of early initiatives of dissemination of science information among the population. In support of both these foci, a number of science organisations were initiated. Important examples include Kerala Sastra Sahitya Parishad in 1962 and its launch of 'science jahtas', in which artists, scientists, teachers and unemployed youth, coordinated by NGOs, travelled to villages with science-based programmes; the People's Science Movement, later the All India's People's Science Network, founded in the 1980s; and the governmental NCSTC (1982). These organisations used a number of different communication processes (media, TV, street theatre, informal lectures, etc.), supported by vast national surveys conducted by institutes such as NCSTC and the National Institute of Science, Technology and Development Studies (NISTAD) of the Council for Scientific and Industrial Research, to underpin a better understanding of the publics by scientists, communicators and government.

In 1982, the Indian government initiated an Indigenous Knowledge Systems National Programme. The aim of this was to support the incorporation of traditional knowledge in national research incentives and to integrate indigenous knowledge systems into the mainstream science system. This led to new directions in the philosophy and sociology of science, by integrating rural communities in the research process. There was a constant drive in India to facilitate communication about local science knowledge as well as to accommodate communication on the development of modern science. Within the broader context of its science and technology policy to support scientific research, India promoted the appropriate diffusion of research, findings and survey outcomes to all strata of the population in an array of activities that have included TV programmes aimed at youth and a yearly Children's Science Congress. Sinha (2002) stated that the Ministry of Science and Technology in India encouraged the application of science and technology for the development of the poorer sections of society, initiated measures to popularise and stimulate a science orientation among the people, and promoted interaction between scientists, engineers and local communities.

NISTAD facilitated dedicated research projects on public understanding of science that provided insight into people's attitude towards science, as well as their level of science understanding. G. Raza, a science communication researcher with NISTAD, proposed the notion that a 'cultural distance' exists between people's everyday experiences (including so-called indigenous knowledge systems) and the knowledge conveyed through formal modern education. Through extensive surveys, his team of researchers collaborated with local youth organisations and other NGOs to establish direct contact with communities, and gained first-hand experience of the cultural structures and embedded knowledge among people. Their main aim was to collect data whereby they could measure the effect of modern education on these people's complexity of thought, with the use of a model designed to effectively measure the cultural distance between thought structures (Raza et al. 2002; Du Plessis and Raza 2004). One of the findings was that the formal system of modern education operated as a strong determinant in shaping cultural structures as well as influencing the worldview of those outside the formal education system (the so-called illiterates) (Du Plessis and Raza 2004).

A crucial factor in the development of science communication in India, however, lies in the cultural tradition. Arguments on epistemology and ethics are already recorded in ancient religious scripts and have recurred time and time again in India's history. The contemporary relevance of the dialogical tradition and the acceptance of heterodoxy is hard to exaggerate (Sen 2005: xi–xiii).

Africa

The African continent, with 54 countries, more than 700 million people and over 500 different languages, is rich in resources and history, but Africans depend mostly on subsistence farming and contribute little to the world's industrial output. The slave trade and colonialism disrupted local development on such a scale that many African countries today are impoverished and culturally deprived. With corruption and bad governance following the liberation in the 1960s of most countries in Africa, many are still struggling to modernise (Makhurane 1998).

Most countries on the African continent still suffer from underdeveloped research infrastructures, insufficient R&D funding and insufficient numbers of researchers. When these countries gained independence from colonial rule, higher education and research had to be built from scratch. This process is ongoing, and progress is often slow. As early as 1950, a Commission for Technical Cooperation in Africa South of the Sahara (CCTA) was established to organise and structure scientific research in Africa. A similar incentive by scientists led to the creation of the Scientific Council for Africa south of the Sahara (CSA), based in Zaire. The organisation stated that 'CCTA, an instrument of African solidarity, is likewise a bridge between Europe's science and Africa's needs. There are other, wider bridges which may carry more traffic, but the one built and maintained by the commission will remain open whatever political fluctuations may occur' (CCTA/CSA/FAMA; cited by Pfister 2005).[2] In 1964 the CCTA was taken over by the Organisation of African Unity.

In the *World Science Report of 1993*, Odhiambo (1993) stated that Africa could not continue to carry the burden of material poverty and scientific illiteracy that overwhelmed its development programmes. He placed responsibility in the hands of African leaders, and blamed their leadership as fractured and anchorless, leading to 'the prospect of a fractured self-image, a fractured society and a fractured future' (Odhiambo 1993: 90).

The development of science and technology in Africa is a crucial and complex part of continental development. Although political independence in Africa contributed to the expansion of local science and technology, scholars regularly debate the existence of pre-European science in Africa (Makhurane 1998). This includes deliberations on the link between African art and African technology, and the implications for science when the traditional technological knowledge systems of women are incorporated into the scientific and public communication systems (Jegede 1998). The dominance of science from the developed world in the African academic sphere (Jegede 1998) is extensively debated. Its influence is considered along with the perceived lack of a broad African contribution to science, caricatured by the slogan popularised by Ki-Zerbo: 'Silence, development in progress' (Mkandawire 2005: 25).

African academics generally agree that the current quest for knowledge is affected by an inherited cultural attitude lacking commitment to the advancement and systematic understanding and documentation of scientific knowledge (Hountondji 1983). African societies made little formal attempt to investigate scientific theories underpinning the technologies they utilised, and did not nurture a disposition to practise sustained inquiries into many areas of their life and thought processes. Successive generations did not substantially expand the lexicon of knowledge, and were seen as using working solutions repetitively. This approach to science as an essentially utilitarian/experiential knowledge system conflicts with the general understanding of science as a systematic and methodical form of inquiry (Hountondji 1997; Makhurane 1998).

At a UNESCO meeting in 1999,[3] the Directors General and Heads of Science and Technology in the Southern African Developmental Community (SADC) region reported on the development of science in their countries. The reports included details such as: absence of a ministry or department of science and technology and a lack of R&D expertise (Botswana); difficulty in monitoring the science and technology system due to a lack of science and technology indicators (Lesotho); a lack of statistics on R&D expenditure (Malawi); no science and technology coordinating mechanism (Mozambique); a low level of human and financial resources for science and technology coupled with a lack of science and technology coordination, a lack of science/maths teachers and a lack of science and technology information (Namibia); and poor dissemination of information related to the impact of research on development linked to a lack of science and technology coordination (Zimbabwe). This general absence of reliable information and statistics on African science and technology makes it difficult to estimate the current state of science on the continent. Contributing factors to this state of affairs include: the brain drain, low priority ranking of R&D by African governments, lack of funds, technological obsolescence, and last but not least, political instability (Pouris 1999).

In research involving the public understanding of science in Africa, it is progressively accepted that the public communication of science and technology must include the communication of cultural information, economics, political and social values and worldviews. It is considered imperative that one should contextualise scientific information according to the specific circumstances and needs of a given community (Sturgis and Allum 2004). Raza et al. (2002) stated that determinants of the thought complexities of communities in developing countries needed more investigation in order to identify the factors that influence individual and group knowledge systems. They argued that the broad cognitive framework (or worldview) in which acquired knowledge is configured is a sociocultural construct, shaped by quotidian or everyday episodes experienced over generations. They proposed that research into knowledge systems should be 'community-centric'[4] (Raza and du Plessis 2002: 59).

The philosopher Kwasi Wiredu (2000) proposed the idea of an African science and technology epistemology as a political issue, whereby 'science and technology in Africa is apt to give the appearance that Africans want to imitate their erstwhile colonisers'. He also called on philosophers to fulfil the role of providing humane rationality and social ethics in response to the anticipated threat of technology for communities lacking in science understanding. Hountondji (1983) posited that Africa will have to develop African science and African scientific research by changing to a literate society (civic literacy). The chief requirement for science is to 'embark on the broad democratic practice of writing' as it is 'difficult to imagine a scientific tradition in a society in which knowledge can be transmitted only orally' (Hountondji 1983: 99–101).

South Africa

South Africa succinctly reflects the complexity involved in the African development of science and technology. Here, public communication of science and technology includes modern as well as traditional technologies. It is a country of contradictions – multiple cultural differences exist in what is often described as a country simultaneously being 'first world' and 'third world'.[5] Governmental inputs to the promotion of science and technology, science communication and indigenous knowledge systems still need improvement in South Africa. Under National Party rule (1948–94), apartheid segregated the local first- and third-world communities through legislation. South African science and technology research was highly sophisticated, even including successful nuclear power research. Such research was highly secretive and well funded, but the imperatives of the black majority were ignored, neglected, and entirely disconnected from the scientific needs of the white minority. Apartheid supported the highly problematic notion of 'race' as an indicator of superiority, and thus compartmentalised its different publics and their cultural knowledge systems (McKinley 1997).

After the first democratic elections in 1994, the African National Congress (ANC) became the ruling party, and was given the task of restructuring this complex and fractured society. The new Minister of Arts, Culture, Science and Technology, Dr Ben Ngubane, declared:

> With the publication of key reports such as the UNDP 2001 World Human Development Report, it is becoming clear that the relationship between science, knowledge and the availability of human capital to address the issues of sustainable development is crucial. This is a very different approach from the

> traditional and narrow thinking of development economics. Few, if any, future scenarios for Africa talk about the contributions of African science and technology to the sustainable development of our planet. This is surely not right. Perhaps we have convinced ourselves that Africa cannot be a player in the knowledge economy. I believe that it is this mindset that needs to be fractured and removed from our consciousness. Science and technology are often seen by policymakers as instruments that have well defined functionality. Under these conditions, science and technology becomes the handmaiden of greater goals such as economic development or quality of life. This instrumentalist approach does great damage because it does not recognise that the potential of people trained in the science and technology is far greater than the primary scientific knowledge that they hold. Scientists and technologists are problem solvers, innovators, entrepreneurs.
>
> (Ngubane 2001)

Science communication was identified as one of the tools for addressing the restrictive practices of the past. South Africa's disparate publics now needed information – but how does one begin to address a lack of science communication among the majority, who also lack rudimentary science education? The ANC proposed that education should be transformed and that teaching of science would create a scientifically literate community. Research and development programmes should be restructured to take into account the neglected field of indigenous knowledge systems.[6] Government support for the development of public understanding of science was spelled out in *South Africa's Green Paper on Science and Technology* (South Africa Ministry of Arts, Culture, Science and Technology 1996). This document identified the need to make information available to society and to recognise the value of equitable public access to information as a tool of transformation.

The South African government proposed to establish a national system of innovation to bring about a society that will understand and value science, engineering and technology, respecting the key concepts and principles of science and technology through knowing that they are social tools, and learning to use science and technology knowledge in ways that enhance personal, social, economic and community development. The 1996 Green Paper noted that 'many campaigns and initiatives have been launched to promote public understanding of science and technology in both developed and developing countries. However, insufficient data are available to assess the outcome of these campaigns'. But it went on to state that 'institutions in South Africa that can play a key role in such initiatives are, for example, societies for the advancement of science, science museums and libraries, media (printed and electronic), educational institutions, private business, government and parliament'. And it proposed that the 'kind of information [made available] be determined by the public who would need to make informed decisions about technology related issues' and that 'the media be identified through which Science and Technology information can be made more accessible to the public' (South Africa Ministry of Arts, Culture, Science and Technology 1996: 84).

A number of government-supported initiatives to promote and develop science and technology have been launched since the mid-1990s:

- African educators and international scholars, under the auspices of the African Forum for Children's Literacy in Science and Technology, participated in December 1995 in a conference to reflect on the continent's needs for science and technology education and to plan future interventions to promote science education.
- The National Research Foundation hosted the Second National Conference on the Public Understanding of Science and Technology in Southern Africa in November 1998.
- The South African Agency for Science and Technology Advancement (SAASTA), which is responsible for a number of science communication activities, hosted the first African Science Communication Conference in December 2006.
- SAASTA co-hosted the world Public Communication of Science and Technology conference in 2002.
- The South African government initiated the National Advisory Council on Innovation, with the first council appointed in 1998. One of its objectives is to promote the public understanding of science and technology and to act in a supportive role in innovation for development and progress. To further these aims, a Foundation for Technological Innovation was proposed in 2007 to bridge the so-called innovation chasm between industry, tertiary education institutions and government departments.

Final remarks

When Amartya Sen (2005) stated that 'the critical voice is the ally of the aggrieved, and participation in arguments is a general opportunity, not a specialised skill', he promoted the idea that the argumentative tradition in India, used with deliberation and commitment, is an extremely important tool in resisting social inequalities and in removing material poverty and intellectual deprivation. With the addition of voice as a crucial accompaniment to the pursuit of social justice, it becomes clear that science communication should not only be seen as a practical tool to support and accompany research development and economic growth. The complexity and diversity of education and communication in science and technology in the developing world turn the local documentation of progress in public communication of science into a daunting task. True debate and participation between scientists and publics are required.

India and Africa are both vast and complex regions, with many of Africa's countries in a constant state of flux. The rules governing scientific research and science education define and constrain what is valued, while producing intellectual legitimacy for their participants (Volmink 1998). In the developed world, intellectual insight in the field of public understanding of science and technology demonstrates a growing awareness that the public communication of science needs to include the lived experience of all societies. The need is therefore expressed for the recognition of value in all ideas and perspectives, including that of indigenous communities who live outside the developed sphere. This need is echoed by academics from Africa, and demonstrated by those in India, but with the clear understanding that science education is needed first before any effective science communication can take place. Hountondji (2002: 132–3) boldly states that freedom of expression and the written word are prerequisites for the development of science in Africa. Future dialogue around the public communication of science and technology should lead to exciting explorations of ideas generated from within the developing world.

Notes

1 Nehru not only took India through the necessary steps of liberation from British colonialism to independence, but introduced far-reaching programmes to help in restructuring his country. He formulated a five-point plan to be implemented within the first few years after the 1947 independence of India:
   - 'People should develop along lines of their own genius and we should avoid imposing anything on them.
   - Tribal rights in land and forest should be protected.
   - We should try to train and build up a team of their own people to do the work of administration and development.
   - We should not over-administer these areas or overwhelm them with a multiplicity of schemes.
   - We should judge results not by statistics or the amount of money spent but by the quality of human life that is involved.' (cited by Pachauri, 1983: 3).

2 Both organisations, due to the internal political situation in South Africa, ironically enough, ceased all relations with South Africa and Portugal in February 1962 (CCTA 1964).

3 The UNESCO meeting had as its topic 'Science and Technology in the SADC Region for the 21st Century', and took place in Pretoria, South Africa, 20–21 April 1999.

4 Raza defined 'community-centric' research as: '. . . a collective which is repository of knowledge that has been generated through process of distillation of abstract ideas extracted from experiential episodes. The spectrum of such communities is quite wide in developing countries. At the one end of the spectrum are those communities which live in harmony with nature without disturbing the regenerative capabilities of ecosystems and who, for example, practise indigenous systems of medicine developed over centuries. On the other hand there are those artisans who have developed what are often referred to as innovative rural or indigenous technologies' (Raza and du Plessis, 2002: 59)

5 These two 'worlds' were part of what was identified during the Cold War period as a tripartite structure of first world (western industrialised, capitalist nations), second world (centralised, command economies in communist countries) and third world (new nations previously colonised by the first world) ideologies. Clear preference was given to the capitalist structure of the developing world, whereby 'the ideological underpinnings of this asymmetric structure politicized the three groups, tainting the transfer of aid and technical assistance with propagandistic overtones' (Margolin 2007: 111).

6 Ngubane stated: 'I am not only talking of what is sometimes wrongly called Western or "first-world" science. South Africa, like many countries, is recognising the unique potential of the knowledge resources of our people. Indigenous knowledge systems, as they have been called, hold great promise in providing a way of lowering the alienation many people feel from science and technology as traditionally taught. Indigenous knowledge projects in South Africa have already shown a rich potential for better curriculum development, as well as new technological innovation' (*Sunday Times*, 11 November 2001).

Suggested further reading

Bulmer, M. and Warwick, D. (1998) *Social Research in Developing Countries: Surveys and Censuses in the Third World*, London: UCL Press.

Du Plessis, H. and Raza, G. (2004) 'Indigenous culture as a knowledge system', *Tydskrif vir Letterkunde* (online), 41, 2: www.ajol.info/viewarticle.php?id=16219
Naidoo, P. and Savage, M. (eds) (1998) *African Science and Technology Education into the New Millennium: Practice, Policy and Priorities*, Kenwyn: Juta & Co.
Raza, G., Singh, S. and Dutt, B. (2002) 'Public, science and cultural distance', *Science Communication*, 23: 293–309.
Riana, D. and Habib, I. (2004) *Domesticating Modern Science*, New Delhi: Tulika Books.

Other references

Bulmer, M. and Warwick, D. (1998) *Social Research in Developing Countries: Surveys and Censuses in the Third World*, London: UCL Press.
CCTA (1964) Publication No. 92 of the CCTA, Lagos: Commission for Technical Cooperation in Africa South of the Sahara.
Cohn, B. (1996) *Colonialism and its Forms of Knowledge. The British in India*, Princeton, NJ: Princeton University Press.
Du Plessis, H. and Raza, G. (2004) 'Indigenous culture as a knowledge system', *Tydskrif vir Letterkunde* (online), 41, 2: www.ajol.info/viewarticle.php?id=16219
Durant, J. (2000) 'Science communication: the challenge in Europe today', paper presented at the *DFID/UNESCO International Workshop on Science Communication*, London, 3–5 July 2000. London: Department for International Development.
Hountondji, P. (1983) *African Philosophy. Myth and Reality*, Bloomington, IN: Indiana University Press.
—— (1997) *Endogenous Knowledge: Research Trails*, Dakar: Codesria.
—— (2002) 'An alienated literature', in Coetzee, P. H. (ed.) *Philosophy from Africa*, Cape Town: Oxford University Press.
Jegede, O. (1998) 'The knowledge base for learning in science and technology education', in Naidoo, P. and Savage, M. (eds) *African Science and Technology Education into the New Millennium: Practice, Policy and Priorities*, Kenwyn: Juta & Co, 151–76.
Kuhn, T. (1962) *The Structure of Scientific Revolutions*, Chicago, IL: University of Chicago Press.
Makhurane, P. (1998) 'The role of science and technology in development', in Naidoo, P. and Savage, M. (eds) *African Science and Technology Education into the New Millennium: Practice, Policy and Priorities*, Kenwyn: Juta & Co, 23–33.
Margolin, V. (2007) 'Design for development: towards a history', *Design Studies*, 28: 111–5.
McKinley, D. (1997) *The ANC and the Liberation Struggle. A Critical Political Biography*, London: Pluto.
Miller, J. (1983) 'Scientific literacy: a conceptual and empirical review', *Daedalus*, 11: 29–48.
—— (1998) 'The measurement of civic scientific literacy' *Public Understanding of Science*, 7: 203–23.
Mkandawire, T. (ed.) (2005) *African Intellectuals. Rethinking Politics, Language, Gender and Development*, Dakar: Codesria.
Naidoo, P. and Savage, M. (eds) (1998) *African Science and Technology Education into the New Millennium: Practice, Policy and Priorities*, Kenwyn: Juta & Co.
Ngubane, B. (2001) 'Science and technology for sustainable development in Africa, UNESCO Seminar, 8 November, France', *Sunday Times*, 11 November: 22.
Odhiambo, T. (1993) *World Science Report of 1993*, Paris: UNESCO.
Pachauri, S. K. (1983) *Dynamics of Rural Development in Tribal Areas*, New Delhi: Concept.
Pattnaik, B. 1992. *The Scientific Temper*, Jaipur: Rawat Publishers.
Pfister, R. (2005) *Apartheid South Africa and African States: From Pariah to Middle Power: 1961–1994*, London: I.B. Tauris.
Pouris, A. (1999) *Science and Technology in the SADC Region for the 21st Century: Executive Summary*, www.unesco.org/science/wcs/meetings/afr_pretoria_99.htm

Raza, G. and du Plessis, H. (eds) (2002) *Science, Craft and Knowledge*, Pretoria: Protea.

Raza, G., Singh, S. and Dutt, B. (2002) 'Public, science and cultural distance', *Science Communication*, 23: 293–309.

Riana, D. and Habib, I. (2004) *Domesticating Modern Science*, New Delhi: Tulika Books.

Sarkar, S. (1975) 'Rammohun Roy and the break with the past', in Joshi, V. C. (ed.) *Rammohun Roy and the Process of Modernisation in India*, New Delhi: Vikas, 45–68.

Sen, A. (2005) *The Argumentative Indian. Writings on Indian Culture, History and Identity*, London: Penguin.

Shen, B. S. P. (1975) 'Science literacy and the public understanding of science', in Day, S. B. (ed.) *Communication of Scientific Information*, Basel: S. Katger AG, 44–52.

Sinha, A. (2002) 'Developing functional capability among artisans: challenges for the trainer', in Raza, G. and du Plessis, H. (eds) *Science, Craft and Knowledge*, Pretoria: Protea.

South Africa Ministry of Arts, Culture, Science and Technology (1996) *South Africa's Green Paper on Science and Technology: Preparing for the 21st Century*, Pretoria: Department of Arts, Culture, Science and Technology.

Sturgis, P. and Allum, N. 2004 'Science in society: re-evaluating the deficit model of public attitudes', *Public Understanding of Science*, 13: 55–74.

Volmink, J. (1998) 'Who shapes the discourse on science and technology education?' in Naidoo, P. and Savage, M. (eds) *African Science and Technology Education into the New Millennium: Practice, Policy and Priorities*, Kenwyn: Juta & Co, 61–78.

Wiredu, K. (2000) 'Our problem of knowledge: brief reflections on knowledge and development in Africa', in Karp, I. and Masolo, D. (eds) *African Philosophy as Cultural Inquiry*, Bloomington, IN: Indiana University Press.

16

Communicating the social sciences

Angela Cassidy

This chapter reviews the sparse and somewhat scattered research literature that has specifically addressed the public communication of the social sciences (PCSS). This literature, in common with much research on the public communication of science and technology (PCST), lacks consistency or clear definitions of what is meant by 'social science', 'natural science' and, indeed, 'science'. Analyses of social science media coverage indicate that the social sciences are communicated in some quite different patterns from those seen with natural science research. Some authors have suggested that this may be due to the overlap between the subject matter of social science research (people) and experiential, 'common-sense' knowledge. This chapter also discusses other relevant literature on 'self-help' psychology books, public intellectuals, and social scientists as expert witnesses.

There is an urgent need for more consistent, systematic research addressing PCSS, in order to understand better the general issues involved in communicating expertise, and those faced specifically by the social sciences. Researchers in PCST should reflect on these issues in order to address reflexively how we communicate publicly about our field, just as we seek to advise other researchers on how best to communicate in the public domain.

A partial view of sciences

As the other contributions to this volume attest, research and practice in many aspects of PCST have grown hugely in recent years. Despite this growth, relatively little attention has been paid to the specific issues that social scientists are faced with when communicating about their research in the public domain. In the published literature, 'science communication' is usually taken to mean the physical, chemical, biological and occasionally medical or engineering sciences, and work on PCSS has been relatively sparse and scattered across several disciplinary areas. In part, this may be because the historical impetus for science communication and public understanding of science has mostly come from natural science. However, the more recent critique of traditional 'deficit model' approaches to science communication has come from social scientists. In this light, it is curious that such researchers have not applied this critique to their own disciplines of sociology, history, anthropology and communication studies, and rarely (if ever) conduct research studies of PCSS. In most countries, mass media have specifically targeted outputs for covering 'science' in one form or another, such as TV and radio programmes about science, science sections in newspapers, and popular science books. Science journalism is a well recognised and respected journalistic specialism, and such professionals provide content for both specialist and mainstream media output. The majority of this coverage tends to be of the above-mentioned natural science disciplines, although social sciences such as psychology do receive some specialist attention. Particularly in the English-speaking media, there is little or no corresponding specialisation in the social sciences, which, as I explore here, has profound implications for the kind of coverage they receive.

The weakness of research in PCSS and of specialisation in social sciences among media professionals is all the more striking when we consider that social science research is covered extensively in the broader, non-specialist media. Crime figures, census data, educational research, economic analysis, psychology and political theory are all examples of social science that contribute to the central, day-to-day content of modern media. Events promoting public discussion of science, such as the annual meeting of the British Association for the Advancement of Science, frequently feature a good deal of social science, showing that it can play a prominent role in science communication at times. The much discussed role of the 'public intellectual' is often taken on by social scientists, particularly in the USA and continental Europe. Social research forms the core activity of many think-tanks, active by definition in the public domain. Social researchers frequently act as expert witnesses in court, and offer policy, personal and lifestyle advice to all and sundry. In the UK, in particular, social scientists have been instrumental in the development of PCST as a research field, and in initiating widespread change in the policy and practice of both governmental and scientific institutions. It is with all this in mind that I review what is currently known about PCSS, and ask the associated question as to why social scientists, particularly those in PCST, have paid so little attention to the popular discussion of their own work.

Literature on PCSS

Research and writing about social science communication is highly disparate and exists in a number of unrelated areas, but when brought together this work provides interesting insights into what happens to social science in the public domain. There are quantitative and qualitative analyses of media content; interview-based studies; analytical work; and material written by social scientists or professional associations addressing the promotion of social science in the public domain, including 'how-to' guides for academics. Of these, the latter is the most commonplace, recalling in many ways the PCST literature of 15–20 years ago. As with much PCST research, the literature is dominated by studies based in the USA and UK, and this bias has important implications for our understanding of social science communication. In English-speaking countries, relatively strong distinctions are drawn between natural science (studies of the natural world) and social science (studies of the human and social) research. This is often associated with a 'hierarchy of the sciences', which places physics at the top, life sciences a little lower, social science below that, and arts/humanities subjects as often outside 'science' altogether. Furthermore, popular ideas about the nature of science reinforce the status of subjects that use quantitative, experimental or statistical methods. In continental Europe and elsewhere in the world, such distinctions are less starkly drawn, and conceptions of science tend to include all forms of scholarly research in their remit, as conceptualised in the German term *Wissenschaft*. Due to this anglocentric bias, as well as my own language constraints, the conclusions drawn in this piece must apply largely to English-speaking countries. Where possible, I refer to studies from the rest of the world, but considering the paucity of the literature as a whole, much more research is needed to reach any coherent understanding of the effect of these cross-cultural differences on PCSS.

Much of the available literature on PCSS is by psychologists; material coming from other disciplines, or addressing social science as a whole, is far less common. This is probably indicative of psychology's borderline status between the natural and social sciences, where the use of quantitative, experimental approaches boosts its 'scientific' status, making it more likely to be covered by journalists (Schmierbach 2005). The greater prominence of psychology may also reflect psychologists' greater concern with their own public image compared with other social sciences, a concern shared with natural sciences, and seen in the more extensive and longer-standing use of media relations by psychological associations and events. Subject area notwithstanding, a great deal of PCSS literature is written by social researchers drawing on their own communication experiences, and frequently resembles older PCST literature in the emphasis on how to get the 'correct' message across in the media (Haslam and Bryman 1994; Kirschner and Kirschner 1997). In this literature, the public image problems of social science are discussed, and strategies to improve the situation often centre on upbraiding journalists for sensationalism, inaccuracy, and lack of understanding of the social science research process (McCall and Stocking 1982; Goldstein 1986). In recent years, professional associations and funding bodies have paid greater attention to improving the skills of researchers when engaging with the media (ESRC 1993, 2005; Gaber 2005). However, the overriding concern remains with the promotion of the social sciences, rather than a more reflective engagement with the issues at hand.

A second area of literature, in places closely related to the above, consists of content-analysis studies of social science media coverage. Weiss and Singer (1988) carried out an extensive study of the American news media during the 1980s, comprising parallel content analysis and interview studies. Key findings included the discovery that the majority of coverage was framed as stories 'about' the subject of the research (for example, crime or parenting), with the research itself appearing as ancillary references. Furthermore, only seven per cent of the stories found were written by specialist science journalists, with most coverage authored by generalists or specialists in other areas. The coverage was analysed by content theme rather than subject area, but it is still plain that economics commanded the largest share of social science coverage in the US media. A similar approach, using a broader sweep of methods, was taken in British research in the following decade (Fenton et al. 1997, 1998), and this study reveals an interesting pattern of similarities and differences between the USA and UK. As in the USA, social science was not covered by science journalists in the UK: in fact only one such example was found in the entire sample studied. Instead, named journalists writing on specific issues produced the majority of the coverage, with the rest contributed by specialists in other areas. In contrast to the US study, social issues provided the largest proportion of the coverage, with economic issues coming a distant second, and psychology was the most frequently represented discipline. Fenton et al. (1997, 1998) report that research did provide the main focus of most stories in the UK, rather than ancillary mentions. Most of the social science coverage analysed appeared as features rather than news articles, and social scientists more often appeared reactively as commentators and advisers on specific issues according to the news agenda, rather than being the principal source of stories.

Both these studies also looked at how much, and where, media coverage of social science appeared, and again transatlantic differences emerge: in the USA, coverage was distributed evenly across all forms of media and levels of reporting found were far higher than in the UK, where coverage was heavily concentrated in the broadsheet (or 'quality') press. However, without meaningful comparisons, it is difficult to draw useful conclusions from these figures: are they high or low, and in what terms? Similarly, it is difficult to distinguish whether many of the issues raised by these studies are specific to the social sciences, or are broader concerns shared in the public communication of all research. A study by Evans (1995) deals with this problem by directly comparing US media coverage of social and natural science research. The study found that of the total sample of research coverage, 36 per cent was of social science subjects, although this is not broken down into disciplinary groupings. The Science Museum Media Monitor (Bauer et al. 1995), one of the largest studies of its kind, took a continental European definition of 'science' as inclusive of the social sciences, and reported a gradual increase in the proportion of social science coverage over the second half of the 20th century, eventually reaching similar levels to that found by Evans. A smaller study carried out by Hansen and Dickinson (1992) found that only 15 per cent of research coverage was of social sciences, but related fields such as market research, 'human interest' and science policy/education were separated out from this, leading to a combined figure of 28 per cent. Overall, these studies suggest that the social sciences provide a substantial proportion of media coverage of research in both the USA and UK, overtaken only by health and biomedicine. By contrast, studies on coverage of the social sciences in the German media have found that they are relatively underrepresented (e.g. Böhme-Durr 1992).

Evans (1995) made some interesting comparisons that chime strongly with the findings of qualitative research addressing PCSS issues, discussed below. For example, social science was much less likely than natural science to appear in newspaper science sections, and more likely to be in general news coverage, confirming the idea that science journalists rarely cover the social sciences. In interviews, Dunwoody

(1986) found that US science journalists typically look down on social science research as less 'scientific', express little interest in it, and regard it as requiring little specialist training to report. Evans (op. cit.) also found that social scientists were accorded a lower epistemological status in media reports, being less often referred to as 'researchers' or 'scientists' and more frequently in terms such as 'the authors of the study'. Finally, he notes the lack of specific 'source' academic journals for media coverage of social science research, compared with major natural science sources such as *Nature* and *Science*. In my study of UK press coverage of evolutionary psychology (Cassidy 2005), I also found that, compared with evolutionary biology, evolutionary psychology was covered less often by science journalists, more by non-specialists, appeared more frequently in features and commentary pieces, and rarely in specialist science coverage. Fenton et al.'s (1997, 1998) research also investigated relationships between social science and the media; they note that social science is not usually covered by correspondents with any in-depth knowledge of research, that it is rarely newsworthy in its own right, and instead is covered as part of changing broader news agendas. Furthermore, they describe the relationship between academics and the media in this area as 'formal, distant and highly reliant on the role of facilitators' (Fenton et al. 1997: 70). Thirty per cent of the researchers they interviewed had worked with the media only via communications professionals, a pattern reflected in interactions between researchers and journalists at academic conferences (ibid.).

Status of social science

Contrary to what might be expected, social science faces many institutional barriers in public communication that are less of a problem for natural science, mostly relating to the relative lack of status accorded to social research in the public domain, and traditional hierarchies of science that place physics and mathematics at their apex and messier, 'subjective' subjects such as sociology and anthropology lower down. Although many social scientists may have abandoned such notions, they remain powerful in popular culture, and have a strong effect on journalistic practice and cultural attitudes. Divisions of labour between science journalists (often trained in natural science) and other journalists (who often have a non-research background, for example in humanities) may well serve to reinforce such notions (Schmierbach 2005). As described above, Anglo-American definitions of 'science' that exclude the social sciences (particularly qualitative research) may accentuate these effects. However, research conducted in Germany suggests that they are more universal: science journalists there also judged social science disciplines negatively, and in general, social science in the media was less prestigious than natural science (Böhme-Durr 1992; see also Wessler 1995).

Although the social sciences may not always enjoy the same status in the media as the natural sciences, it may be that social scientists play a different, and at times more influential, role as experts in the wider public domain. A Norwegian study (Kyvik 2005) found that academics in the humanities and social sciences were more publicly active (in terms of writing popular articles and media contributions such as availability for interview) than their colleagues in natural science, medical and technical subjects. This is also suggested by the small body of work on popular and self-help psychology, which is also largely US-based, unsurprising considering the importance of the genre in that country. Many of these studies take a discursive approach, addressing, for example, the regulation of heterosexuality in John Gray's *Mars and Venus* series of books on relationships between men and women (Potts 1998); or masculinity in popular books on child development (Anderson and Accomando 2002). The rhetorical language of pop psychology has been examined, showing how ideas about self-help tie into discourses of individuality and personal growth, and are closely related to New Age movements (Askehave 2004) as well as the broader values of modern liberal democracies (Hazleden 2003). Considering the obvious popularity of these texts, as seen through the vast sales figures they secure not only in the USA but globally, this work gives an insight into an arena where social science is highly influential on ordinary people's lives. Crawford (2004) takes this further by investigating audience responses to the *Mars and Venus* series, analysing a television programme showing couples discussing the book *Men are from Mars, Women are from Venus* (Gray 1992). Crawford argues that this shows how audiences' responses to these books rarely consist of straightforward absorption, and that instead people can use such texts to open up a space for negotiating and challenging the claims made therein.

Another relevant literature is that surrounding the idea of the 'public intellectual' – broadly understood as a person of learning, not necessarily an academic, who uses their knowledge to engage in wider society through debating in the public domain (Small 2002). Although this idea is hardly new, academic and popular discussions of 'the public intellectual' have burgeoned in the past few years. However, this has been a conversation strongly centred on humanities/social sciences, and has barely featured in the literature on PCST issues. Although many people thought of as public intellectuals, such as the late Edward Said (1994), have been writers and thinkers in the humanities, many others are and have been social scientists. A survey about public intellectuals, carried out by the UK political magazine *Prospect* in 2005, put a social scientist (Noam Chomsky) at the top of its list of 100 public intellectuals, and many other social scientists featured prominently. Therefore the literature on public intellectuals is of interest to anyone concerned with PCSS. A key debate in this literature turns on the 'duty' of public intellectuals to be politically engaged in society, and how they should best carry this out (Alcoff 2002). In a similar vein, the sociologist Michael Burawoy (2005) has called for 'public sociology', arguing that this is a role that should be taken on more by his colleagues. This has sparked debates about the role of sociology in the public domain and whether it should be politically engaged at all (Clawson et al. 2007). A recent paper on this touched on the potential risks and pitfalls of engaging with the media, describing a specific case, how the research message was changed by media reporting, and the lack of media awareness of the authors (Grauerholtz and Baker-Sperry 2007). This suggests that both areas could benefit from some cross-talk: while recent PCST research has had much to say about academic engagement with the media, broader issues of expertise, politics and the public domain have received less attention (but see Chapter 9 in this volume).

The case of psychology

Psychoanalysis is also an important area for understanding the popular influences of social science, particularly in terms of the above literature on self-help books. Serge Moscovici's (1961) classic study of popular psychoanalysis provides an unusually historical and European view of PCSS issues, addressing three different parts of French society during the 1950s. Moscovici argued that each milieu carried a slightly different 'social representation' of psychoanalytical ideas, each of which reflected its own values and ideas. More recently, Park (2004) has compared the contemporary discourses of popular psychiatrists with those of psychoanalysts, arguing that the two groups strategically position themselves against each other, respectively as medical, 'scientific' experts, and as broader intellectual authorities. He relates these opposing, yet complementary strategies to the differing forms of public intellectual visible in contemporary popular culture. Unlike the popular scientist, the public intellectual comments on a broad range of issues, rather than keeping to their own area of expertise. It could be argued that natural scientists such as Richard Dawkins and Stephen Jay Gould have taken on such a role. This may signal an increasing source of 'competition' for social scientists' expertise in the public domain, as natural science research on subjects such as genetics and neuroscience comments increasingly on traditionally 'social science' research topics.

Research on UK media coverage of the newly emerging subject of evolutionary psychology has also investigated these tensions, as evolutionary psychology is located on the boundaries between the natural and social sciences. Most evolutionary psychologists work within positivist, naturalistic and quantitative traditions of social science, such as cognitive psychology. Through a series of popular books, public lectures, media interviews and articles, they argued in favour of using evolutionary theory and quantitative, experimental methods to research human behaviour, society and culture. At the same time, these popular arguments attacked opposing theories and methods for understanding 'human nature', such as the interpretive research traditions of sociology and anthropology, and cultural explanations of human behaviour. This popular coverage helped evolutionary psychologists to reach audiences across disciplinary boundaries, and move the subject from a relatively marginalised position to one of establishment in academia (Cassidy 2005, 2006). It also stimulated a public debate in the UK media, involving psychologists, philosophers, biologists, feminists, novelists, commentators and journalists. This tied into other issues under debate at the time, including heterosexual relationships, feminism and gender, centre-left politics, and the prominence and role of bioscience (such as genetics, evolution and neurobiology) in understanding and governing society (Cassidy 2007). This research also picked out a key issue for understanding PCSS – the overlap between the expert knowledge of social science researchers and people's everyday experience of human existence. Participation in evolutionary psychology debates was not restricted to accredited 'experts' in the subject, but instead included a much wider range of people. Like much social science, it was rarely covered as 'science' in the media, and 'lay' forms of knowledge such as personal experience, common sense and gender identity were often drawn upon to make arguments both for and against evolutionary psychology (Cassidy 2004).

Paying attention to the subject matter of social science research is central to understanding how it is communicated and understood in the public domain. Because the social sciences investigate the realm of the human – people, their minds, societies, money, politics, and so on – the subjects, researchers, communicators and audiences of research tend to bleed into one another. Unlike the natural sciences, where expertise is almost by definition held by researchers and specialists, social scientists' expertise is often about matters of everyday experience and common-sense knowledge, affecting how highly that expertise is regarded. For example, Evans (1995) reports that US journalists made strong demarcations between natural science and social science, between natural science and lay opinion, but not between social science and lay opinion. As psychologists McCall and Stocking (1982: 988) put it:

> Everyone, including journalists and editors, fancies himself or herself something of a psychologist, but not an astrophysicist. Results from psychology, but not physics, must therefore square with experience to be credible.

Fenton et al. (1998) found that news media audiences do precisely this in framing their understandings of social science research findings. They also found that the consequent overlap between the professional role of the social scientist and that of the journalist resulted in further underreporting of social research, as journalists often felt it was little different from their own work, and therefore not inherently newsworthy. Similar issues of the 'scientific' legitimacy of social science expertise have also been seen in studies of social scientists' role as expert witnesses. Legal definitions of 'science' in the USA are often heavily traditional, positivist ones, leading at times to non-natural science expertise being judged as inadmissible (Lynch and Cole 2005).

However, these overlaps between social science, journalism and everyday knowledge also have positive implications for PCSS. Historian of psychology Graham Richards (2002) has described this phenomenon as 'reflexive science', while Fenton et al. (1998: 102) refer to it as 'epistemological consonance'. Media 'news values', which often result in natural science struggling to gain media coverage, can often work in favour of the social sciences. Examples of such news values include relevance (to daily life), consonance (with existing beliefs), topicality, controversy, and of course human/personal interest (Weiss and Singer 1988: 144–9; Fenton et al. 1998: 103–13; Gregory and Miller 1998: 110–4). In this light, it becomes less surprising that the content analyses discussed above showed social science to be a very widely reported area of research. Epistemological consonance can also help explain media attitudes that journalists do not require specialist training to report social science, ironically also increasing the chances of social science research being reported in the first place. As described above, generalists tend not to have training in either natural or social science, and neither do editors, ironically increasing the chances that social science will make it through the editorial process. Furthermore, social scientists can use, and have used, the overlap between their research and everyday knowledge to help popularise their work, emphasising the commonality or separation between the two areas, according to their rhetorical purposes (Derksen 1997, 2000; Shapin 2001).

Concluding remarks

Although the literature on PCSS is sparse and scattered across many disciplinary areas, some interesting trends have emerged, alongside striking gaps in the literature and opportunities for further research, many pertaining specifically to the social sciences. A close examination of social science communication also opens up some crucial questions for the broader field of PCST research. With so little work done, it is difficult to reach firm conclusions about social science communication, so any assertions made here are by necessity highly provisional and subject to further investigation. Despite this, one thing is immediately clear: social science is simultaneously marginalised and immensely popular in the public domain, at least in the English-speaking world. Social science research has a lower epistemological status than natural science, is less likely to be newsworthy in and of itself, does not merit media or journalistic specialisation, and at times is seen as little different from journalism itself. Much of this stems from the social sciences' marginal status on the boundaries of 'science', both in mass media and wider society. At the same time, social science is very frequently covered by the media, seen as relevant to audiences, easy to understand, and appears throughout media coverage rather than being confined to an area of special interest, as with natural science. As such, social scientists often have important roles to play as commentators and advisers on social, political and personal issues.

Beyond these rather broad-brush assertions, it is difficult at this stage to draw any more nuanced conclusions about social science communication. The criteria used for coding content-analysis studies have been so variable that it is very difficult to draw meaningful comparisons between them. They have been carried out in different countries, sometimes decades apart, over different timescales, coding for different data and using variable definitions of social science and, indeed, 'science', which is also a problem for PCST research in general. Therefore the most urgent need is for further work, preferably using a comparative approach, to look at the communication of a broad spread of disciplines, including social sciences. In the studies reviewed here, some intriguing suggestions of cross-cultural differences have emerged, particularly in the popularity of particular disciplines, and of the social sciences as a whole. However, little work has been done outside the Anglo-American context, so further studies, in continental Europe and in the rest of the world, are also urgently needed. Similarly, little attention has been paid in PCST research to the role of historical context and change in PCSS. This is notwithstanding the 40-year period under study in the Bauer et al. (1995) study of the British press, and a thriving literature addressing histories of popular (natural) science (Cantor and Shuttleworth 2004).

Although much research attention has been paid to the work of specialist science journalists, the widespread reporting of social science by non-specialists highlights the fact that little or no work has been done on how generalists understand and report academic research. The reflexive nature of social science research, and the idea that this is what makes PCSS so different from PCST, is one that also requires further investigation. This is particularly as it may also cast light on what makes communicating natural science so difficult at times, particularly in those subjects very far from human experience. Furthermore, it may provide an important contribution to current debates about the construction of expertise (Collins and Evans 2002) and the related issues of public engagement with science and scientific decision-making (Leach et al. 2005). Understanding how and why PCSS is different from PCST would also be helpful for practitioners, as communicating the complex findings of social science, particularly those of qualitative research, presents a significant challenge. As noted above, social scientists have until recently paid far less attention to the public and communicative aspects of their work than their colleagues in the natural sciences. Considering that social science research is far more likely to be of relevance and importance to public debates, media reporting and, indeed, the majority of ordinary people, there has been a curious lack of attention among social researchers to these issues.

As a final note, I would like to present a challenge to researchers and practitioners working in PCST: how do we communicate about our work on communication, and publicly engage about public engagement? Surely, if we aim to advise natural scientists, policy-makers and politicians about these issues, then we should practise what we preach, and communicate openly and ably ourselves. However, the extra levels of reflexivity introduced in PCST work ('communicating about research which is about communicating about research') is unlikely to be compatible with media news values, for example. Surely one of the most urgent challenges facing our field is to start looking for the answers to such questions, both through further research work and by providing practical examples of engaging in the public domain about the importance of PCST research.

Suggested further reading.

Crawford, M. (2004) 'Mars and Venus collide: a discursive analysis of marital self-help psychology', *Feminism and Psychology,* 14: 63–79.

Evans, W. (1995) 'The mundane and the arcane: prestige media coverage of social and natural science', *Journalism and Mass Communication Quarterly,* 72: 168–77.

Fenton, N., Bryman, A., Deacon, D. and Birmingham, P. (1998) *Mediating Social Science*, London: Sage.

Lynch, M. and Cole, S. (2005) 'Science and technology studies on trial: dilemmas of expertise', *Social Studies of Science,* 35: 269–311.

Small, E. (2002) (ed.) *The Public Intellectual*, London: Blackwell.

Weiss, C. H. and Singer, E. (1988) *Reporting of Social Science in the National Media*, New York: Russell Sage Foundation.

Other References.

Alcoff, L. M. (2002) 'Does the public intellectual have intellectual integrity?', *Metaphilosophy,* 33: 521–34.

Anderson, K. J. and Accomando, C. (2002) 'Real boys? Manufacturing masculinity and erasing privilege in popular books on raising boys', *Feminism and Psychology,* 12: 491–516.

Askehave, I. (2004) 'If language is a game – these are the rules: a search into the rhetoric of the spiritual self-help book *If Life is a Game – These are the Rules*', *Discourse and Society,* 15: 5–31.

Bauer, M., Durant, J., Ragnarsdottir, A. and Rudolfsdottir, A. (1995) *Science and Technology in the British Press 1946–1990: A Systematic Content Analysis of the Press (Vols I–IV)*, London: Science Museum.

Böhme-Durr, K. (1992) 'Social and natural sciences in German periodicals', *Communications: The European Journal of Communication Research*, 17: 167–76.
Burawoy, M. (2005) '2004 Presidential Address: For Public Sociology', *American Sociological Review*, 70: 4–28.
Cantor, G. and Shuttleworth, S. (eds) (2004) *Science Serialised: Representations of the Sciences in Nineteenth Century Periodicals*, Cambridge, MA: MIT Press.
Cassidy, A. (2004) 'Of academics, publishers and journalists: evolutionary psychology in the UK media', PhD thesis, Edinburgh: University of Edinburgh.
—— (2005) 'Popular evolutionary psychology in the UK: an unusual case of science in the media?', *Public Understanding of Science*, 14: 115–41.
—— (2006) 'Evolutionary psychology as public science and boundary work', *Public Understanding of Science*, 15: 175–205.
—— (2007) 'The (sexual) politics of evolution: popular controversy in the late twentieth century UK', *History of Psychology*, 10: 199–227.
Clawson, D. Zussman, R., Misra, J., Gerstel, N., Stokes, R., Anderton, D. and Burawoy, M. (2007) *Public Sociology: Fifteen Eminent Sociologists Debate Politics and the Profession in the Twenty-first Century*, Berkeley, CA: University of California Press.
Collins, H. M. and Evans, R. (2002) 'The third wave of science studies: studies of expertise and experience', *Social Studies of Science*, 32: 235–96.
Crawford, M. (2004) 'Mars and Venus collide: a discursive analysis of marital self-help psychology', *Feminism and Psychology*, 14: 63–79.
Derksen, M. (1997) 'Are we not experimenting then? The rhetorical demarcation of psychology and common sense', *Theory and Psychology*, 7: 435–56.
—— (2000) 'Boundaries and commonplaces: the rhetorical demarcation of common sense, paper presented at Demarcation Socialised: How Can We Recognise Science When We See It?' conference organised by the Centre for Knowledge, Expertise and Science, Cardiff University, 24–27 August 2000.
Dunwoody, S. (1986) 'The science writing inner club: a communication link between science and the lay public', in Friedman, S. L., Dunwoody, S. and Rogers, C. L. (eds) *Scientists and Journalists: Reporting Science as News*, New York: Macmillan.
ESRC (1993) *Pressing Home Your Findings: Media Guidelines for ESRC Researchers*, Swindon: Economic and Social Research Council.
—— (2005) *Communications Toolkit*, Swindon: Economic and Social Research Council. www.esrc.ac.uk/ESRCInfoCentre/Support/Communications%5FToolkit
Evans, W. (1995) 'The mundane and the arcane: prestige media coverage of social and natural science', *Journalism and Mass Communication Quarterly*, 72: 168–77.
Fenton, N., Bryman, A., Deacon, D. and Birmingham, P. (1997) 'Sod off and find us a boffin: journalists and the social science research process', *Sociological Review*, 45: 1–23.
Fenton, N., Bryman, A., Deacon, D. and Birmingham, P. (1998) *Mediating Social Science*, London: Sage.
Gaber, A. (2005) 'Media coverage of sociology', *Sociological Research Online*, 10(3), www.socresonline.org.uk/10/3/gaber.html
Goldstein, J. H. (ed.) (1986) *Reporting Science: The Case of Aggression*, Hillsdale, NJ: Lawrence Erlbaum Associates.
Grauerholtz, L. and Baker-Sperry, L. (2007) 'Feminist research in the public domain: risks and recommendations', *Gender and Society*, 21: 272–94.
Gray, J. (1992) *Men are from Mars, Women are from Venus: A Practical Guide for Improving Communication and Getting What You Want in Your Relationships*, New York: HarperCollins.
Gregory, J. and Miller, S. (1998) *Science in Public: Communication, Culture and Credibility*, New York: Plenum Trade.

Hansen, A. and Dickinson, R. (1992) 'Science coverage in the British mass media: media output and source input', *Communications*, 17(3), 365–77

Haslam, C. and Bryman, A. (1994) *Social Scientists Meet the Media*, London: Routledge.

Hazleden, R. (2003) 'Love yourself: the relationship of the self with itself in popular self-help texts', *Journal of Sociology*, 39: 413–28.

Kirschner, S. and Kirschner, D. A. (1997) *Perspectives on Psychology and the Media*, Washington, DC: American Psychological Association.

Kyvik, S. (2005) 'Popular science publishing and contributions to public discourse among university faculty', *Communications*, 26, 2885–311.

Leach, M., Scoones, I. and Wynne, B. (2005) *Science and Citizens: Globalisation and the Challenge of Engagement*, London: Zed Books.

Lynch, M. and Cole, S. (2005) 'Science and technology studies on trial: dilemmas of expertise', *Social Studies of Science*, 35: 269–311.

McCall, R. S. and Stocking, S. H. (1982) 'Between scientists and public: communicating. psychological research through the mass media', *American Psychologist*, 37: 985–95.

Moscovici, S. (1961) *La Psychanalyse: Son Image et son* Public, Paris: Presses Universitaires de France.

Park, D. W. (2004) 'The couch and the clinic: the cultural authority of popular psychiatry and psychoanalysis', *Cultural Studies*, 18: 109–33.

Potts, A. (1998) 'The science/fiction of sex: John Gray's *Mars and Venus in the Bedroom*', *Sexualities*, 1: 153–73.

Richards, G. (2002) *Putting Psychology in its Place: A Critical Historical Overview* (2nd edn), London: Routledge.

Said, E. W. (1994) *Representations of the Intellectual: The 1993 Reith Lectures*, London: Vintage.

Schmierbach, M. (2005) 'Method matters: the influence of methodology on journalists assessments of social science research', *Science Communication*, 26: 269–87.

Shapin, S. (2001) 'Proverbial economies: how an understanding of some linguistic and social features of common sense can throw some light on more prestigious bodies of knowledge', *Social Studies of Science*, 31: 731–69.

Small, E. (2002) (ed.) *The Public Intellectual*, London: Blackwell.

Weiss, C. H. and Singer, E. (1988) *Reporting of Social Science in the National Media*, New York: Russell Sage Foundation.

Wessler, H. (1995) 'Die journalistische Verwendung sozialwissenschaftlichen Wissens und ihre Bedeutung für gesellschaftliche Diskurse [The journalistic use of social-scientific knowledge and its relevance for social discourses]', *Publizistik*, 40: 20–38.

17

Evaluating public communication of science and technology

Federico Neresini and Giuseppe Pellegrini

In its everyday usage, the term 'evaluation' is employed in a wide variety of contexts to denote the act of expressing judgement on some activity or other. This is a natural exercise, which seems to require neither careful thought nor particular skills: we evaluate when we decide whether the lunch we have just had was enjoyable, whether a colleague has done his or her work well, whether the person to whom we are talking has understood what we want to say.

Besides the variety of objects to which it is applied, evaluation therefore means establishing the extent, at least approximately, to which a given action has produced effects that match the purposes for which it was undertaken.

This definition is still valid when evaluation leaves the everyday domain and becomes a set of activities performed to assess the results of actions that are more complex and more structured – as well as more ambitious – than those of everyday routine. We thus have evaluation of educational performance, the social impact of a particular political decision, the quality of a public service, the claimed advantages of a technological application, the effects of a communication campaign.

In this perspective, evaluation becomes a structured and formal activity, a systematic inquiry that applies specific procedures in gathering and analysing information on the content, structure and results of a project, programme or planned intervention (Guba and Lincoln 1989). Consequently, evaluation may also perform a political role in supporting decision-making processes and choices on programmes and activities by trying to reduce the level of uncertainty (Patton 1986: 14). The task of evaluation, therefore, is to yield systematic evidence that informs experience and judgement, furnishing an array of options available to the actors involved in a programme (Weiss 1990: 83).

In short, evaluation relates to what determines or explains the success or failure of an action in regard to the goals for which it was first conceived and then undertaken.

What does it mean to evaluate PCST activities?

Such a general definition is of little help in understanding the practice of evaluation more precisely, or for defining the particular characteristics of the evaluation of an activity in public communication of science and technology (PCST). In respect of the first point, it could be useful to consider how the practice of evaluation corresponds to a specific area of social research; answering this question could give us a useful ground from which to examine the second.

Evaluation as a particular kind of social research

It is no coincidence that evaluation and social research share a number of key terms: for instance analysing, understanding, measuring, explaining. Evaluation, in fact, is nothing other than social research applied for the purposes just mentioned.

Evaluation must therefore deal with the epistemological and methodological issues that concern social research in general. It would be beyond the scope of this article to examine these issues in detail. Nevertheless, they should be discussed briefly in order to highlight certain aspects of particular importance for evaluative research.

Consider, for example, 'GM Nation?', an initiative promoted by the British government between 2002 and 2003 and involving activities of various kinds – preliminary workshops, a series of different kinds of public meeting, a dedicated website, focus groups (actually termed 'narrow-but-deep' groups) and a survey for collecting participants' feedback (36,553 completed questionnaires) – intended 'to promote an innovative and effective programme of public debate on issues around GM in agriculture and the environment, in the context of the possible commercial growing of GM crops in the UK' (PDSB 2003: 11). Although the initiative's principal purpose was to involve the public in important decisions concerning the regulation of biotechnologies in the agri-food sector, it also had communicative goals, in that it also sought to provide the public with the information that it needed to participate in the debate.

The initiative required a large commitment of resources (approximately €1 million), and when it ended, debate began on evaluation of its results. The discussion covered numerous aspects, among them the ability of 'GM Nation?' to generate real participation, to give voice to all the positions present in society, and to furnish guidelines for legislation (Barbagallo and Nelson 2005; Horlick-Jones et al. 2006; Irwin 2006). A number of issues arose in the course of the debate, for example: What is meant by 'participation'? Who are the 'public'? How can 'representativeness' be defined? How can it be established whether, and to what extent, the initiative has been successful? It is evident that these are questions with a crucial bearing on the aims pursued by 'GM Nation?'

The fact that no agreement had been reached on these questions on conclusion of the initiative highlights the lack of a sufficiently clear *ex ante* definition of its objectives. It is obvious that the actors involved in 'GM Nation?' pursued different goals, or at any rate gave different meanings to the terms used to formulate them (Rowe et al. 2005). In other words, each actor observed – and therefore evaluated – the process and its results from a different point of view.

A criterion with which to solve the point of view problem once and for all obviously does not exist, for every solution has its pros and cons. But precisely for this reason, although it is still the result of choices and negotiations, evaluative research should be able to rely on some reference parameters that make it possible to establish which perspective has been adopted. In short, evaluation produces results that have value only in relation to the context in which they have been obtained – starting from the aims of the project being observed – rather than in absolute terms.

Another methodological problem that evaluation shares with social research has to do with the opposition between so-called quantitative methods – principally surveys with standardised questionnaires – and qualitative ones (discursive interviews, ethnographic observation, focus groups, etc.). The former methods tend to favour a type of interaction between the researcher and the phenomenon observed, which is characterised by detachment, neutrality and separation; the latter emphasise involvement, a direct relation, and a sort of constitutive reciprocity.[1] Nevertheless, if we look at the evaluative research already done – both in the PCST area and in general – we can clearly see that making a rigid opposition between quantitative and qualitative research methods is not very useful in practice.

As an example, the researchers who, albeit as outsiders, evaluated 'GM Nation?' adopted a 'multi-method approach that used both quantitative and qualitative methods' (Pidgeon et al. 2005).[2] Using different tools is a good strategy for improving the evaluation's appropriateness with regard to the nature and the context of the communication programme (Joubert 2007). In the case of the Robot Thought Project, for example, an interactive performance was assessed using observation, exit survey and video analysis (Graphic Science UWE 2004). The specific nature and the aims of the project required a wide assortment of tools to make data collection and data processing easier; these made it possible to study different reactions to the performance, assessing the process of communication.

The experience of 'GM Nation?' is also very interesting because the inclusion of external evaluators *in itinere* – in the sense that it was not envisaged at the outset – highlighted a further problem typical of social research, generally referred to as 'access to the field'. As in all social research, the presence of researchers in the field must be legitimated and therefore previously negotiated, so evaluative research must be recognised and accepted by the actors about to be observed. But in the case of 'GM Nation?' 'there was no indication of any enthusiasm on the part of government to undertake such an evaluation itself' (Grant 2003). Moreover, 'the late presence of evaluators also raised concerns with the recruited contractors and sub-contractors, who clearly felt a degree of disquiet at suddenly finding out that their activities would be under scrutiny, leading to some difficult negotiations amongst the various parties' (Pidgeon et al. 2005). If evaluation is not envisaged and planned from the outset, with adequate resources made available, it will not be carried out unless additional resources unexpectedly become available. We are doubtless looking at a useless but decisive consideration: as with any other programme of social research, evaluation needs resources to be put into practice.

Is there any specificity in evaluating PCST initiatives ?

Evaluative research's dependence on the context of its realisation – dependence on the objectives and on the type of activity involved in reaching these objectives –

raises questions as to whether, on the one hand, evaluating PCST activities is different from evaluating other activities and, on the other, if there is a relationship between how PCST is interpreted and the evaluation of the activity that follows it. There does appear to be a connection between the conception and evaluation of a PCST activity: evaluating a PCST activity with the purpose of transmitting knowledge is not the same as evaluating another activity with the purpose of promoting discussion between different social actors about a certain issue. In the first case, we are dealing with top-down communicative interaction; in the second, communication is based mainly on dialogue. Also, in the first case the results are largely predetermined and in the second the PCST activity will be more open-ended. The move from activities that focus on transmitting knowledge to activities oriented towards dialogue and discussion obviously means a passage – well described in this volume – from the 'deficit model' perspective to that of involvement and participation (see Chapters 5 and 14 in this volume). If PCST activities with different aims can be carried out using tools that result in attitude, behaviour and knowledge changes, their evaluation should also be able to take this into account. As we will see later, there are many different methodologies available; their selection depends on the type of transformation expected as a result of a PCST activity. Evaluating the quantity and quality of the changes produced in terms of knowledge entails methodological choices and evaluation techniques that are different from the evaluation of changes in attitudes or behaviour, but this does not necessarily correspond to different initial aims.

By focusing attention on who will be the 'object' of the evaluation, it is possible to find a connection between the different aims of a PCST initiative, the different kinds of communication promoted through it, and its evaluation. An activity devoted mainly to transferring knowledge or to persuading the public will be evaluated in terms of the changes produced at this level. If the aim is also to promote dialogue or participation, the promoters of the communication initiative will have to be under observation.

This happened, for example, with the BIOPOP project in which young European biotechnologists met citizens trying to develop innovative models of communication of life science, realised through two public events in Bologna (Italy) and Delft (the Netherlands). Under a tent placed in the main square of these two cities, young researchers talked with people about biotechnologies, starting from very simple activities such as looking into a microscope, doing PC games and moulding biological molecules by hand (www.biopop-eu.org). As the project's aims were focused on dialogue, the evaluation had to take into account not only effects on the public participating in the events organised, but also effects on the young biotechnologists conducting the project.

Evaluation of a series of consensus conferences organised in the Netherlands in 1994–95 tried to assess, on the one hand, the impact on participants – with respect to knowledge and attitude – and, on the other, that on policy-makers (Mayer et al. 1995). In this perspective, evaluation makes it possible also to collect evidence of unexpected effects, as in the case of the 'citizen conference on genetic diagnostics' carried out in Dresden by the German Hygiene Museum in 2001. In fact, 'the results were generally positive, though policymakers and scientists were dissatisfied with the outcome of the conference: participants grew more critical toward some of the diagnostic techniques rather than more accepting' (Storksdieck and Falk 2004: 100–1). Evaluation that is aimed not only at the participants therefore becomes a good indicator for the difference between PCST activities that are deficit model-oriented and those more interested in dialogue and participation (Table 17.1).

It can be seen that the more PCST activities aim at dialogue and participation, the more their evaluation tends to coincide with that of initiatives aimed at engaging the public in decision-making on issues with a high techno-scientific content. Even if we are dealing with a wide range of different activities, they often present interesting challenges in evaluation focused not only on the public side (Joss and Durant 1995; Rowe et al. 2005).

Evaluating the effects of communication

The organisation of a communication campaign, the mounting of an exhibition, the media launch of a news story, the construction of a website, the creation of a science centre: these, and all other activities related to PCST, develop through time. Like other human activities, they can be divided into three phases: design (*ex-ante*), implementation (*in itinere*) and conclusion (*ex-post*). Although this is an obvious oversimplification, this distinction is nevertheless a useful means to focus on certain crucial aspects of evaluation.

Focusing on the design phase, evaluation concentrates on the adequacy of the resources available with respect to the objectives to be pursued. To this end, assessment is made not only of financial and time aspects, and of human resources, but also of whether the communicative strategy adopted will be able to reach the target audience. Knowledge of the principal features of the interlocutors to be involved in the communication is therefore crucial (Storksdieck and Falk 2004); and it may also be important for subsequent evaluation of the results.[3]

Evaluation *in itinere* corresponds largely to 'formative evaluation' – evaluation intended to establish what is working and what is not in the ongoing activity, and to adjust the latter accordingly (Scriven 1991). This requires analysis of patterns of interaction among the actors, identification of obstacles and unforeseen effects, and monitoring of the use to which the available resources are put. Evaluation *in itinere* often also uses content analysis and ethnographic observation (for an overview of the potential and application of the various content-analysis techniques see Gaskell 2000; Bauer and Gaskell 2001). The evaluation focus can be, for example, on what happens at a public meeting between scientists and members of the public – as in the case of a

Table 17.1 PCST initiatives and their evaluation

PCST activities and goals	*Type of communicative interaction*	*Type of expected outcome*	*Whose effects to be observed?*
Information	Top-down	Highly predetermined	Public
Convincing/Persuading	Top-down	Highly predetermined	Public
Discussion/Dialogue	Horizontal	Mainly open-ended	All social actors involved

'consensus conference'[4] – or how a visit to a museum or a science centre develops as an experience involving not only learning, but also entertainment, social relationships and emotions (Kotler and Kotler 1998; Storksdieck and Falk 2004).[5] Information is also collected during an ongoing communicative process to conduct the 'theory-based evaluation' proposed by Weiss. This is 'a mode of evaluation that brings to the surface the underlying assumptions about why a programme will work. It then tracks those assumptions through the collection and the analysis of data at a series of stages along the way to final outcomes. The evaluation then follows each step to see whether the events assumed to take place in the program actually take place' (Weiss 2001: 103; see also Gascoigne and Metcalfe 2001).

A good example of formative evaluation can be found in the Large Hadron Collider (LHC) Communication Project promoted in 2006 by the British Particle Physics and Astronomy Research Council. The LHC project seeks 'to engage the public with particle physics, developing a four-year programme with the twin aims of increasing public knowledge of, and support for, particle physics and inspiring young people to choose physics courses at 16 and subsequent decision points' (PSP, 2006). The evaluation was carried out to assess levels of knowledge and understanding of both particle physics and the basic scientific questions that are driving the LHC project, public perception, and relevant questions to improve the communication process. These objectives were studied for two target groups: members of the general public who are interested in science, and students and their teachers. Formative evaluation, conducted through focus groups, interviews, questionnaires and discussions, gave useful results that helped refine the methods and content of communication in the course of the project.

Evaluation finds its natural place at the end of the communicative process, because it aims to determine and explain the success or failure of an action with respect to the goals it was initially intended to achieve (summative evaluation). In the case of the Stavanger bus campaign, for example, people were interviewed to assess the efficacy of a campaign of advertisements with science messages on city buses for a target audience of 16–35-year-olds (Graphic Science UWE 2003).[6] The campaign was evaluated using a quantitative survey with the aim of collecting data directly from the target population; the key findings allowed the communication tools used to be assessed.

It is usual in this regard to distinguish between output and outcome. In the former case, the results are defined as the effective accomplishment of what the initial design envisaged, thus privileging the point of view of the promoters of the communication. In the latter case, the results are instead viewed as changes produced by the communication, so that the attention focuses – at least potentially – on all the actors involved in the process. Evaluation may yield contrasting judgements in the two cases but, even more importantly, good results in terms of output offer no guarantees in terms of outcome. For this reason, we concentrate here mainly on evaluation of the effects of PCST activities.

Types of change involved in communication

Communication has paradoxical features, also displayed by other aspects of everyday life when examined from within the social sciences. Although communication is a practice constitutive of social relations, as well as of individual experience, and although it has been the subject of numerous studies, it is still a rather obscure phenomenon: for example, a generally agreed definition of communication has not yet been produced. (In the specific case of PCST, see the overview by Bucchi 2004.) However communication is defined, there is some agreement that it is a process able to engender change in those who take part in it (Watzlawick et al. 1967; Bateson 1972; Maturana and Varela 1980; Von Foerster 1981). This raises an intriguing question: when can we say that communication has come about? The answer is apparently rather simple: when we can say that something has changed for those involved in the communication process.

This is particularly important as regards evaluation. For if communication produces change, then the aim of evaluation must be to establish the extent and nature of this change. In other words, the intrinsic capacity of communication to produce change in those who take part in it makes understanding its effects of special relevance, as well as interest. It is not by chance that the problems of the effects of communication have received close attention from scholars of the mass media: 'If any one issue can be said to have motivated media studies, it is the question of "effects"' (Jensen 2002: 138). All researchers interested in social interaction – from the micro- to the macro-level – have sooner or later had to confront this problem.

If there is to be some hope of success in evaluating the effects of communication, one must have an idea – general and therefore somewhat imprecise – of what changes communication can plausibly be expected to produce. Assembling suggestions from a variety of sources, but all of relevance to PCST, we may say that such changes may take place at the level of knowledge, attitudes or behaviour.

Distinguishing among these three types of change is anything but straightforward. It may be rather obvious that a change at the level of knowledge has to do with learning (I now know that a molecule of water consists of two hydrogen atoms and one oxygen atom); at the same time, it is clear that altering the way we conceive a given aspect of our experience – in substance, changing the way we express judgements – pertains to the level of attitudes (contrary to what I thought before, I now believe that scientists are reliable), and that if I begin to do something I did not do before (watch science programmes on television, attend a science café, encourage my son or daughter to enrol on a science degree course), this concerns the level of behaviour. However, learning cannot be reduced to the acquisition of information alone; it also concerns change in our interpretive schemas or cognitive models, meaning the criteria on which we base our judgements, pertaining to learning. Expressing a judgement is, likewise, a form of behaviour and involves a certain body of knowledge; while performing an action involves a motivational dimension made up of beliefs, competences and knowledge (Michael 2002).

Although these qualifications raise serious doubts as to the validity of the above classification, distinguishing changes at the level of knowledge, attitudes and behaviour is still useful when addressing the evaluation of effects of communication.

Detecting and understanding these three types of change raise somewhat diverse problems from the point of view of the methodology of social research. Rather than seeking the optimal solution, it is important to be aware of the pros and cons of each social research technique available for this purpose. For example, while a change consisting in the acquisition of information can be measured by means of a standardised questionnaire, change in interpretive patterns could be better observed by using more flexible techniques, such as the in-depth interview. Ethnographic observation may be better suited to documenting the behaviour of visitors to a science centre, perhaps using the shadowing technique (Sachs 1993; Fletcher 1999). But if we are interested in a change in attitudes, the focus group seems to offer greater advantages.

However, there are various aspects that should be borne in mind. For example, the rigidity of the standardised questionnaire, together with the well-known dependency of respondents on the context in which the questionnaire is submitted and on the wording of questions, may introduce considerable bias into the data used for evaluation; but the development of a discursive interview depends largely on the characteristics of the interviewer and of the setting in which the interview takes place. The data collected through focus groups may reflect, among other things, the way in which the discussion group has been composed. Such considerations also arise in ethnographic research.

The central problem: observing change

Whether the concern is to transmit information or to promote interaction on an equal footing between scientists and the public, the principal problem of evaluation is still the same: how can the change produced by a communicative process be observed? The apparently simplest and most obvious solution goes by the name of 'experimental design'. Put briefly, the approach consists of comparing the situation *ex-ante* the communicative event with the situation *ex post*, assuming that any changes observed are due to the communication that has taken place. Taking this approach, one group has to be involved in a communication process (for example, viewing a TV programme on science) and another group has not; if the two groups have been made up through a randomised selection, their possible differences observed *ex-post* are attributed to the communication experience. The same result can be obtained by comparing certain characteristics of, say, a group representing the target audience of an information campaign or the visitors to a science centre with those of a control group not involved in the communicative process. Here, too, any changes may be attributed to the communication.

Although attractive at first sight, the experimental design solution proves in reality to be fraught with difficulties. Many of these difficulties derive from the impossibility of complying with the requirement underlying every kind of experimental – or quasi-experimental[7] – research, namely the 'other things being equal' (*ceteris paribus*) condition. For example, if it is decided to compare the knowledge – or the attitudes or the behaviour – of the visitors to a science museum before and after their visits, the very activity of *ex-ante* data collection will produce changes in the subjects, maybe by prompting them to pay closer attention to certain contents of the exhibition than they would do otherwise. For this reason, what one observes as the visitors leave the museum will not be the difference between what the subjects knew before and what they know now due to their involvement in the communicative process stimulated by the exhibition, but rather the difference between what the subjects knew before and what they know now due to their involvement in a communicative process stimulated by the exhibition *and* the questions put to them on entering. The reason for this 'slippage' is very simple: the administration of a questionnaire or the conduct of an interview are also communicative processes and, as such, they produce a change in the participants.

But if one seeks to deal with this difficulty by comparing with a control group in a quasi-experimental design, there arises another, almost insoluble, problem of social research: the impossibility of obtaining two groups that are completely identical or, in the more attenuated version of the problem, groups that are sufficiently similar, where 'sufficiently' means 'equal in the characteristics most relevant to the changes that one wants to observe'. Visiting the website of one scientific institution rather than another may not have a great deal to do with attitudes towards science, or with knowledge of evolutionary theory, but being pestered by parents to get good marks at school most certainly does, even if one might not think of collecting information about it.

The problem also concerns more refined methodological solutions, such as those depicted in Table 17.2 as 'quasi-experimental' design with 'crossing groups', for which the difficulty of obtaining homogeneity between the groups compared is not resolved. According to this research design, the target audience of an initiative in science communication – for example, the students at a particular school – in whom changes in characteristics A and B are expected to occur, are divided into two homogeneous groups consisting of students attending the fourth year (alpha and beta). The situation with regard to characteristics A and B is analysed before the event begins, but group alpha only for A and group beta only for B. Then, after the event, the situation of characteristic B is surveyed for the alpha group, and the situation of A for the beta group. It is thus possible to compare the characteristics surveyed *ex ante* and *ex post* with respect to A of two different but homogeneous groups, thereby reducing the problem of the conditioning exerted by the *ex-ante* data collection on the *ex-post* one. However, in this case too, the requirement of perfect homogeneity between the two groups compared is still difficult to fulfil.[8]

Once again, the optimal solution does not seem to exist, even if one abandons the (quasi-)experimental approach and adopts what we may call '*ex-post* observation'. Rather than comparing the *ex-ante* and the *ex-post* situations, the researcher can now

Table 17.2 Evaluation and experimental design

Approach	*Problems*
(Quasi-)experimental *ex-ante/ex-post* design	The *ex-ante* data-collection activity conditions the *ex-post* survey
(Quasi-)experimental design with a control group	Homogeneity between the experimental group and the control group
(Quasi-)experimental design with crossing groups	Homogeneity between the groups compared
Ex-post (self-)evaluation by subjects	Self-deception and deference to interviewers
Ex-post observation	Strong hypotheses are necessary on the characteristics of the groups segmented according to the different behaviours observed

examine only the characteristics of the *ex-post* one, interpreting them as indicators of changes produced by the communication. In this case, the target audience – for example, teachers attending an in-service training course – performs self-evaluation of communication initiatives by means of both standardised questionnaires and in-depth interviews. In both cases, however, there are potential problems of self-deception and of deference to the interviewers. Moreover, the researcher is obliged to proceed without the benchmark (the *ex-ante* situation) necessary to determine any changes that may have occurred.

Alternatively, the researcher may try to survey the change indicators indirectly, and then verify whether they assume different values in various segments of the sample defined by variables that relate to the change expected. For example, following a communication campaign, the researcher could test indicators of interest in nano-technologies, and then determine whether their values differ significantly between subjects with low and high levels of schooling, if other evidence has shown that education level is a discriminating variable with respect to interest in scientific research. If the lower-educated group continues to show little interest in science, while the higher-educated group is still very interested, one may conclude that the communication campaign has not produced major changes, or that, at best, the changes have simply reinforced already existing attitudes. In this case, the main limitation consists in the need to possess quite detailed knowledge about the relations between certain characteristics of the target audience and other characteristics that the campaign is intended to modify.

We should not underestimate the difficulty arising from the fact that the majority of communicative events are of brief duration, while we often expect them to produce great changes with respect to knowledge, attitudes or behaviour. In other words, there is an evident disproportion between a meeting of citizens and scientists, even if it lasts for an entire day, and the acquisition of new knowledge, new interpretive patterns and new habits. There is an even greater disproportion when the communicative process consists of viewing a television programme, reading an article, or visiting a science centre. Evaluation should take account of this aspect (Storksdieck and Falk 2004) and awareness of the importance of the 'time factor' has led to the development of various research strategies. Among the best known of these are the following:

- analysis of short-term effects, bearing in mind that they are highly unstable (after I have visited an exhibition on quarks, I can remember a great many things, but how many will I remember some months later?);
- study of the effects generated by repeated involvement in numerous communicative events of the same type (this is the idea at the basis of numerous research studies on the long-term effects produced by the mass media);
- research designs that envisage at least one follow-up survey to determine which of the changes recorded in the short period are sufficiently consolidated to be still observable some time later.

Also with regard to the time factor, the technical solutions that can be adopted for identifying and analysing change differ in their nature, first, because these solutions vary according to the purposes for which communication processes have been initiated; second, because each of them has strengths and weaknesses.

The main concern in this discussion is not to identify a social research method for use in evaluation without contra-indications. What evaluation requires is adequate awareness of the advantages and disadvantages of the methodology adopted to observe the change generated by communication. In the end, what matters is not isolating the effects of communication, on the one hand, and its causes, on the other; rather, it is understanding the ways in which they combine. If one seeks to find a non-existent optimal solution, there is a serious risk that final evaluation will be abandoned because it is believed to be impracticable.

In this regard, distinguishing between 'sequential causation' and 'generative causation' may be particularly useful. In the former case – following Pawson and Tilley (1997) – the evaluation seeks to verify whether a particular result can be legitimately ascribed to a given input, assuming that the cause–effect relation thus identified can be generalised to other situations. In the latter case, it is instead important to understand how the observed result may have ensued from a given input, emphasising the role that the context – in combination with the input – plays in producing the result.

Evaluation of attendance at an open day held by a scientific institution will not be interested in establishing a causal nexus between what the visitors have seen and any changes in their knowledge, attitudes or behaviour. Rather, it will be interested in relating the type of communicative experience made possible by that context to the effects observed, seeking to take account of the inevitable shortcomings of such observation. It may even be found that the activity of observation itself, in that it is part of the overall communicative process experienced during the open day, has contributed directly to producing the effects ascertained.

Final remarks

Despite its limitations, difficulties, and perhaps even its contradictions, the evaluation of the public communication of science is still useful and necessary.

As we have already stressed, evaluation implies that the objectives of the public communication of science be made clear, with specification – especially by its promoters – of the results that a proposed initiative can be legitimately expected to produce. This means bringing to the surface the assumptions that tend to remain hidden behind what we take for granted, so that their implications can be scrutinised.

It is important to bear in mind that these assumptions are necessary because they constitute the motivational basis of commitment in PCST activities, and unverifiable because they cannot be demonstrated to be true or false, only regarded as more or less convincing. On this view, the assumptions of the deficit model (knowledge generates attitudes, and these determine behaviour) can be easily identified in the background of numerous communication initiatives, even if their declared objectives state otherwise, and they can be subjected to more detailed scrutiny. But this is not to say that if these criticisms imply a conviction that, for instance, public communication of science should not seek to transmit knowledge, but instead to stimulate interest in science, this new assumption does not have its own shortcomings, or is absolutely better than the deficit model. Nor will evaluation settle the matter once and for all. Just as happens to science in public debate on science-based issues (Von Schomberg 1995; Grove-White et al. 2000), evaluation cannot cut the Gordian knot tying truth and falsehood together, but rather can heighten our awareness of the effects, often paradoxical (Boudon 1977, 1995), of our activities.

This is not a constraint, but an opportunity – another good reason to put the logic of evaluation into practice, for evaluating is nothing other than learning from experience more systematically and more efficiently than is done spontaneously.

Including adequate evaluation in PCST initiatives will highlight both their strengths and weaknesses, not only by reducing the risk of persevering in the wrong direction, but also by refining our understanding of what those initiatives seek to achieve (Gammon and Burch 2006). PCST tends to proceed by imitation, following the fashion of the moment, neglecting that rapid proliferation of similar initiatives does not necessarily indicate success, and that what has worked in one context will not necessarily produce the same results in another. Evaluation may prove very useful in insulating against the allure of simple replication.

Clarifying the boundaries between credible expectations and unfounded hopes, reflecting carefully on what has been achieved, scrutinising our initial assumptions – only these can ensure that the present euphoria concerning the public communication of science will not be followed by depression due to unfulfilled expectations.

Notes

1 As a consequence, the survey tools used by quantitative methods are therefore more standardised; qualitative methods, by contrast, have greater margins of flexibility, which enables them to adapt to the situation at hand. See, for example, Strauss and Corbin (1998); Silverman (2004).

2 They justified this choice by referring explicitly to the orientations emerging from the methodological debate on evaluation (Rossi et al. 1999; Shaw 1999), even though there were doubts and troubles, as mentioned by Pidgeon et al. (2005); Reynolds and Szerszynki (2006). For specific remarks on PCST, see Bauer et al. (2007).

3 In this regard, useful guidelines have been proposed by Eng et al. (1999). These guidelines are partial adaptations of those set out by the National Cancer Institute in *Making Health Communication Programs Work* (National Cancer Institute 1989).

4 See, for example, Barnes (1999); Mayer et al. (1995). The general implications of evaluation of participation initiatives are discussed by Rowe and Frewer (2004).

5 For a survey see Persson (2000); Piscitelli and Anderson (2000); Garnett (2002). On the relevance of the context in which the communication process takes place see, among others, Schiele and Koster (1998); Falk and Dierking (2000).

6 The Stavanger bus campaign was commissioned by the British Council (Norway) in January 2003. Findings from Stavanger will be set alongside findings from the SciBus project, which ran across all 15 EU member states.

7 A design becomes 'quasi-experimental' when the selection of subjects to be included in the experimental group and in the control group is not randomized.

8 For an example of application of this research design, see the concluding report of the Inside the Big Black Box (IN3B) Project, www.observa.it

Suggested further reading

Gammon, B. and Burch, A. (2006) 'A guide for successfully evaluating science engagement events', in Turney, J. (ed.) *Engaging Science. Thoughts, Deeds, Analysis and Action*, London: Wellcome Trust, 80–5.

Rowe, G., Horlick-Jones, T., Walls, J. and Pidgeon, N. (2005) 'Difficulties in evaluating public engagement initiatives: reflections on an evaluation of the UK GM Nation? public debate about transgenic crops', *Public Understanding of Science*, 14: 331–52.

Storksdieck, M. and Falk, J. H. (2004) 'Evaluating public understanding of research projects and initiatives', in Chittenden, D., Farmelo, G. and Lewenstein, B. (eds) *Creating Connections*, Walnut Creek, CA: AltaMira Press, 87–108.

Other references

Barbagallo, F. and Nelson, J. (2005) 'Report: UK GM dialogue', *Science Communication*, 26: 318–25.

Barnes, M. (1999) *Building a Deliberative Democracy: An Evaluation of Two Citizens' Juries*, London: Institute for Public Policy Research.

Bateson, G. (1972) *Steps to an Ecology of Mind*, New York: Ballantine Books.

Bauer, M. W. and Gaskell, G. (2001) *Qualitative Researching with Text, Image and Sounds. A Practical Handbook*, London: Sage.

Bauer, M. W., Allum, N. and Miller, S. (2007) 'What can we learn from 25 years of PUS survey research? Liberating and expanding the agenda', *Public Understanding of Science*, 16: 79–95.

Boudon, R. (1977) *Effets pervers et ordre social*, Paris: PUF.

—— (1995) *Le juste et le vrai. Études sur l'objectivité des valeurs et de la connaissance*, Paris: Fayard.

Bucchi, M. (2004) 'Can genetics help us rethink communication? Public communication of science as a "double helix"', *New Genetics and Society*, 23: 269–83.

Eng, T. R., Gustafson, D. H., Henderson, J., Jimison, H. and Patrick, K. (1999) 'Introduction of evaluation of interactive health communication applications', *American Journal of Preventive Medicine*, 16: 10–14.

Falk, J. H. and Dierking, L. D. (2000) *Learning from Museums*, Walnut Creek, CA: AltaMira Press.

Fletcher, J. K. (1999) *Disappearing Acts. Gender, Power and Relational Practice at Work*, Cambridge, MA: MIT Press.

Gammon, B. and Burch, A. (2006) 'A guide for successfully evaluating science engagement events', in Turney, J. (ed.) *Engaging Science. Thoughts, Deeds, Analysis and Action*, London: Wellcome Trust, 80–5.

Garnett, R. (2002) *The Impact of Science Centers/Museums on their Surrounding Communities: Summary Report*, Canberra, Australia: Questacon, www.astc.org/resource/case/Impact_Study02.pdf

Gascoigne, T. and Metcalfe, J. (2001) 'The evaluation of national programs of science awareness', *Science Communication*, 23: 66–76.

Gaskell, G. (eds) (2000) *Qualitative Researching with Text, Image and Sound: A Practical Handbook*, London: Sage.

Grant, M. (2003) *AEBC Meeting*, 11–12 December, Eden Project, Cornwall, London: Agriculture and Environment Biotechnology Commission.

Graphic Science UWE (2003) *Stavanger Bus Campaign Evaluation*, Bristol: University of the West of England, www.uwe.ac.uk

—— (2004) *Robot Thought Evaluation Report*, Bristol: University of the West of England, www.uwe.ac.uk

Grove-White, R., Macnaghten, P. and Wynne, B. (2000) *Wising Up: The Public and New Technologies*, Lancaster: Centre for the Study of Environmental Change, Lancaster University.

Guba, E. and Lincoln, Y. (1989) *Fourth Generation Evaluation*, Newbury Park, CA: Sage.

Horlick-Jones, T., Walls, J., Rowe, G., Pidgeon, N., Poortinga, W. and O'Riordan, T. (2006) 'On evaluating the GM Nation? public debate about the commercialisation of transgenic crops in Britain', *New Genetics and Society*, 25: 265–88.

Irwin, A. (2006) 'Public deliberation and governance: engaging with science and technology in contemporary Europe', *Minerva*, 44: 167–84.

Jensen, K. B. (2002) *The Qualitative Research Process. A Handbook of Media and Communication Research: Qualitative and Quantitative Methodologies*, London: Routledge.

Joss, S. and Durant, J. (1995) *Public Participation in Science: The Role of Consensus Conferences in Europe*, London: Science Museum.

Joubert, M. (2007) 'Evaluating science communication projects', www.scidev.net/ms/sci_comm

Kotler, N. and Kotler, P. (1998) *Museum Strategy and Marketing. Designing Missions, Building Audiences, Generating Revenues and Resource*, San Francisco, CA: Jossey Bass.

Mayer, I., de Vries J. and Guerts, J. (1995) 'An evaluation of the effects of participation in a consensus conference', in Joss, S. and Durant, J. (eds), *Public Participation in Science: The Role of Consensus Conferences in Europe*, London: Science Museum, 201–23.

Maturana, H. and Varela, F. (1980) *Autopoiesis and Cognition. The Realization of Living*, Dordrecht: Riedel.

Michael, M. (2002) 'Comprehension, apprehension, prehension: heterogeneity and the public understanding of science', *Science, Technology and Human Values*, 27: 357–78.

National Cancer Institute (1989) *Making Health Communication Programs Work*, NIH Publication No. 89-1493, Bethesda, MD: National Institutes of Health, US Department of Health and Human Services.

Pawson, R. and Tilley, N. (1997) *Realistic Evaluation*, London: Sage.

Patton, M. Q. (1986) *Utilization-focused Evaluation*, Newbury Park, CA: Sage.

PDSB (2003) *GM Nation? The Findings of the Public Debate. Final Report of the GM Public Debate Steering Board*, London: Clarity.

Persson, P. E. (2000) 'Science centers are thriving and going strong', *Public Understanding of Science*, 9: 449–60.

Pidgeon, N. F., Poortinga, W., Rowe, G., Horlick-Jones, T., Walls, J. and O'Riordan, T. (2005) 'Using surveys in public participation processes for risk decision making: the case of the 2003 British GM Nation? public debate', *Risk Analysis*, 25: 467–79.

Piscitelli, B. and Anderson, D. (2000) 'Young children's learning in museums settings', *Visitor Studies*, 3: 3–10.

PSP (2006) *Formative Evaluation of the LHC Communication Project*, London: People Science & Policy, www.peoplescienceandpolicy.com/projects/hadron_collider.php

Reynolds, L. and Szerszynski, B. (2006) Representing GM Nation? in *Proceedings of the Participatory Approaches in Science & Technology (PATH) Conference, 4–7 June 2006, Edinburgh, UK*, Aberdeen: Macaulay Institute, www.macaulay.ac.uk/PATHconference

Rossi, P. H., Freeman, H. E. and Lipsey, M. W. (1999) *Evaluation. A Systematic Approach*, 6th edn, Thousand Oaks, CA: Sage.

Rowe, G. and Frewer, L. J. (2004) 'Evaluating public-participation exercises: a research agenda', *Science Technology & Human Values*, 29: 512–57.

Rowe, G., Horlick-Jones, T., Walls, J. and Pidgeon, N. (2005) 'Difficulties in evaluating public engagement initiatives: reflections on an evaluation of the UK GM Nation? public debate about transgenic crops', *Public Understanding of Science*, 14: 331–52.

Sachs, P. (1993) 'Shadows in the soup: conceptions of work and nature of evidence', *Quarterly Newsletter of the Laboratory of Human Cognition*, 15: 125–32.

Schiele, B. and Koster, E. (eds) (1998) *La revolution de la muséologie des sciences: vers le musée du XXIe siècle?*, Lyon, France: Presses universitaires de Lyon.

Scriven, M. (1991) *Evaluation Thesaurus*, 4th edn, Thousand Oaks, CA: Sage.

Shaw, I. (1999) *Qualitative Evaluation*, Thousand Oaks, CA: Sage.

Silverman, D. (2004) *Qualitative Research. Theory, Method and Practice*, 2nd edn, London: Sage.

Storksdieck, M. and Falk, J. H. (2004) 'Evaluating public understanding of research projects and initiatives', in Chittenden, D., Farmelo, G. and Lewenstein, B. (eds) *Creating Connections*, Walnut Creek, CA: AltaMira Press, 87–108.

Strauss, A. L. and Corbin, J. M. (1998) *Basics of Qualitative Research: Techniques and Procedures for Developing Grounded Theory*, 2nd edn, Thousand Oaks, CA: Sage.

Von Foerster, H. (1981) *Observing Systems: Selected Papers of Heinz von Foerster*, Seaside, CA: Intersystems Publications.

Von Schomberg, R. (1995) *Contested Technology. Ethics, Risks and Public Debate*, Tilburg: International Centre for Human and Public Affairs.

Watzlawick, P., Beavin, J. H. and Jackson, D. D. (1967) *Pragmatics of Human Communication. A Study of Interactional Patterns, Pathologies and Paradoxes*, New York: Norton & Co.

Weiss, C. H. (1990) 'New directions for program evaluation', in Alkin, M. C., Patton, M. Q. and Weiss, C. H. (eds) *Debates on Evaluation*, Newbury Park, CA: Sage.

—— (2001) 'Theory-based evaluation: theories of change for poverty reduction programs', in Feinstein, O. and Picciotto, R. (eds) *Evaluation and Poverty Reduction*, Washington, DC: World Bank.

Index